Student Solutions Manual
to accompany

Statistics and Probabilty in Modern Life

sixth edition

Joseph Newmark

College of Staten Island
of the City University of New York

SAUNDERS GOLDEN SUNBURST SERIES

Saunders College Publishing
Harcourt Brace College Publishers

Fort Worth Philadelphia San Diego New York Orlando Austin
San Antonio Toronto Montreal London Sydney Tokyo

Newmark: Student Solutions Manual to accompany *Statistics and
Probability in Modern Life, 6e.*

ISBN 0-03-019487-3

7 095 98765432

Preface

This Student Solutions Manual is a supplement for Newmark's *Statistics and Probability for Modern Life*, sixth edition. It contains detailed solutions to every other odd numbered question in the text (including exercises and chapter tests). If you have any comments or suggestions about this Student Solutions Manual, please address your correspondence to: Mathematics Editor, Saunders College Publishing, 150 South Independence Mall West, Public Ledger Building, Suite 1250, Philadelphia, PA 19106.

TABLE OF CONTENTS

CHAPTER 1

ANSWERS TO EXERCISES FOR SECTION 1.4

1. Although it should be descriptive only, in reality it is used as both descriptive and inferential statistics

5. a) The part that indicates that the number of Americans over 65 years of age increased by 6.3% over last year at this time.

    b) The part that predicts that by the end of the century approximately 25% of the American population will be over 65 years of age.

9. It should be descriptive only.

13. Descriptive statistics

21. No. The 10% raise in salary is based on a lower salary.

25. Technically yes, but I would not recommend it.

ANSWERS TO EXERCISES FOR SECTION 1.5

1-6.

ROW	JUNE	SALES	TOTAL
1	1	157802	1002919
2	2	181362	
3	3	159576	
4	4	172887	
5	5	193216	
6	6	138076	

ANSWERS TO EXERCISES FOR SECTION 1.5

9. This command adds the entries in column C2 and places the sum in C3.

13. This command saves the data file as ASSIGN1 on a floppy disk in drive A.

ANSWERS TO TESTING YOUR UNDERSTANDING OF THIS CHAPTER'S CONCEPTS - (PAGE 28)

1. Probably

5. The tobacco industry claims that people may develop lung cancer because of other factors such as air pollution, water pollution, etc.

ANSWERS TO CHAPTER TEST - (PAGES 28 - 32)

1. Choice (c)

5. Choice (d)

9. **a)** Statistical inference   **b)** Descriptive statistics

    c) Statistical inference   **d)** Statistical inference

13. The part that estimates that about 15% of the homes in the region are contaminated with radon gas.

ANSWERS TO CHAPTER TEST - (PAGES 28 - 32)

17. A discrete random variable is a variable that represents observations or measurements of something that results from a count. A continuous random variable is a variable that represents observations or measurements of something that result from a measure of quantity.

ANSWERS TO THINKING CRITICALLY - (PAGES 32 - 34)

1. It is possible that the number of dolphins in the sea has already been reduced from earlier catches.

5. No. Less cars on the road.

9. No. There was a general increase in population.

ANSWERS TO CASE STUDY - (PAGES 34 - 35)

1. and 2. The claims are unsubstantiated.

CHAPTER 2

ANSWERS TO EXERCISES FOR SECTION 2.2

1.

Miles per gallon	Tally	Frequency
25-26	│	1
27-28	││││	4
29-30	⋕ ││││	9
31-32	⋕ ⋕ │	11
33-34	⋕ ││││	9
35-36	⋕ │	6
37-38	⋕ │	6
39-40	││	2
41-42	│	1
43-44	│	1
		50

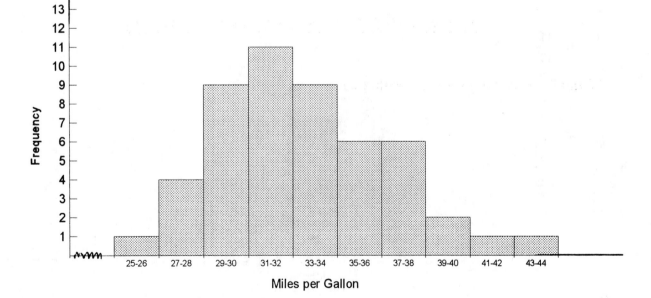

5

5.

Monthly Bill	Tally	Frequency
$14.69-$15.70	|||	3
15.71-16.72			1
16.73-17.74	||||	5
17.75-18.76	|||| |||	8
18.77-19.78	|||| ||	7
19.79-20.80	|||| ||||	9
20.81-21.82	||||	4
21.83-22.84			1
22.85-23.86			1
23.87-24.88			<u>1</u>
		40

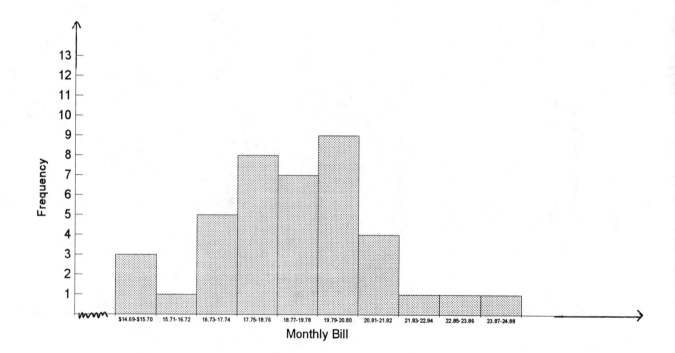

ANSWERS TO EXERCISES FOR SECTION 2.2

9. a)

Number of Products Completed	Tally	Frequency
11-13	|	1
14-16	||||	4
17-19	||||	4
20-22	||||	4
23-25	|	1
26-28	|||	3
29-31	|	1
32-34	||	<u>2</u>
		20

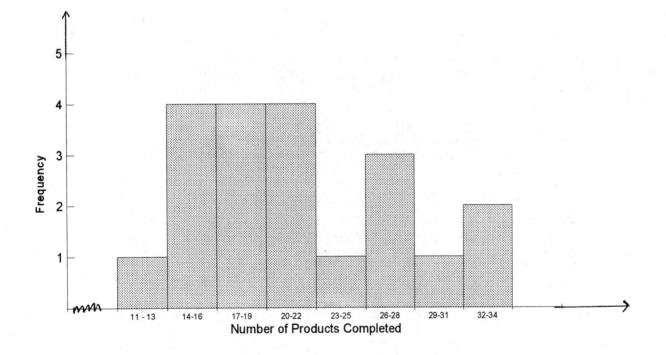

13.

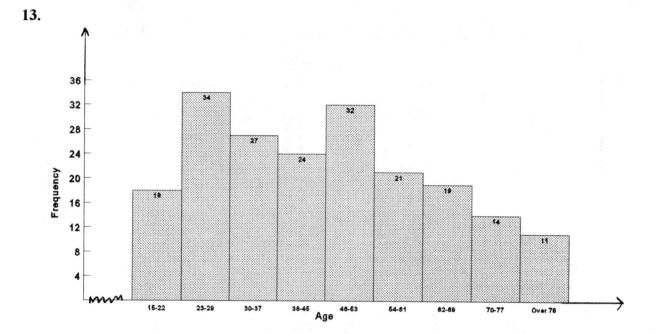

a) $\dfrac{103}{200} = 51.5\%$ b) $\dfrac{121}{200} = 60.5\%$ c) $\dfrac{83}{200} = 41.5\%$

d) $\dfrac{14}{200} = \dfrac{7}{100} = 7\%$ e) $\dfrac{186}{200} = \dfrac{93}{100} = 93\%$

1.

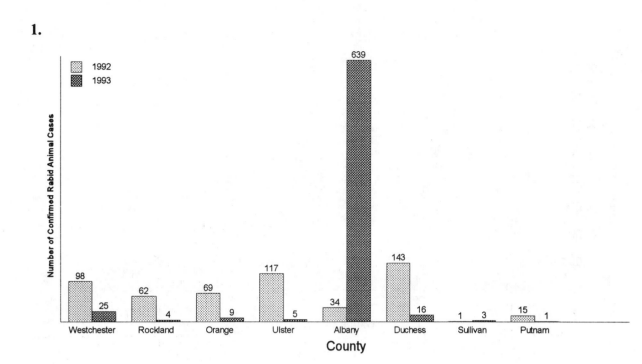

5. a)

Device	1980	1990
Telephone Service	82%	84%
Television	85%	98%
Radio Sets	85%	85%
Cable TV	18%	62%
VCR	2%	77%

b) From 2% to 77% or by 75%

c) No change

9.

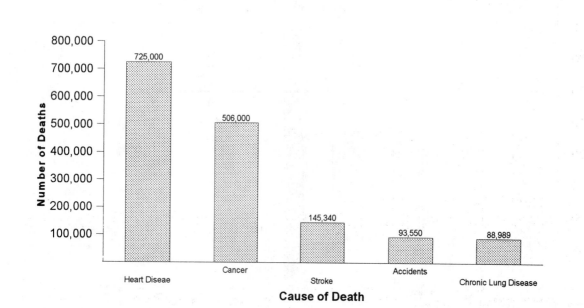

13. a)

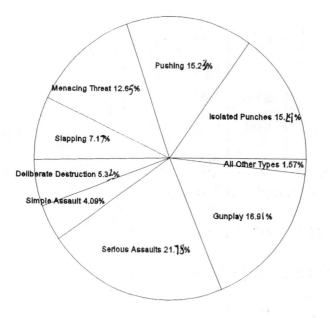

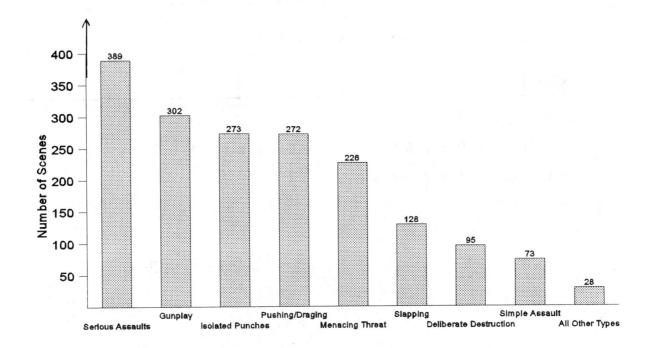

b) Probably the bar graph

17. a) China   : 850 million
  India   : 625 million
  Soviet Union : 250 million
  United States : 225 million

 b) 850 - 250 = 600 million

21. a) Assuming that the 1158 samples were selected from all of the 6 counties together, we have 1158/6 = 193 samples per county.

  Nassau : 0.015 × 193 = 2.895 samples
  Rockland : 0 × 193 = 0
  Suffolk : 0.026 × 193 = 5.018 samples
  Westchester : 0 × 193 = 0
  New York City : 0.013 × 193 = 2.509 samples

 b) The bars of the bar graph are not drawn to scale. For example, consider the height

of the rectangles for Rockland and Westchester (0%).

ANSWERS TO EXERCISES FOR SECTION 2.4

1. a)

Stem	Leaves
2	4 3 4 4 4 4
2	9 6 6 9 5 7 8 9 7 7 7 6 8 7 7 8 8 6 5 8 6 9 7 8 8 9 6 7 8
3	1 0 1 1 2 2 0 3 3 0 0 1 4 2 3

b)

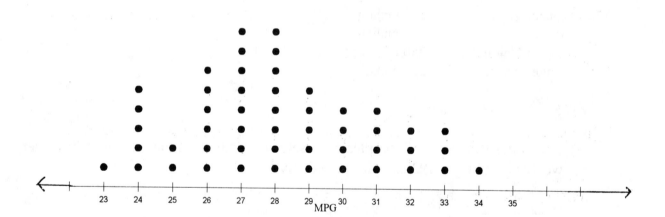

ANSWERS TO EXERCISES FOR SECTION 2.4

5.

Stem	Leaves
18 (1850-1899)	75
19 (1900-1949)	45 45 01 45
19 (1950-1999)	98 89 76 87 98 84
20 (2000-2049)	00 00 04 49
20 (2050-2099)	50
21 (2100-2149)	49 49 45 00
21 (2150-2199)	98 50
22 (2200-2249)	49 49 00

ANSWERS TO EXERCISES FOR SECTION 2.5

1. Both graphs are statistically correct, but because different spacings are used on the vertical scale, the graphs appear different.

5. It should be $\frac{26.6}{19.3}$ or 1.378 times as tall, i.e. different spacings are used on the vertical scale. The conclusion may be true but not based upon the information obtained from this graph.

ANSWERS TO EXERCISES FOR SECTION 2.6

1. $\dfrac{1.38}{1.26} \times 100 = 109.52$ Price increased by approximately 9.52% when

    compared with 1989.

5.

Year	Cost Index	Interpretation
1990	$\dfrac{3300}{3300} \times 100 = 100$	Base year price
1991	$\dfrac{3700}{3300} \times 100 = 112.12$	Price increased by approximately 12.12% when compared with 1990
1992	$\dfrac{4000}{3300} \times 100 = 121.21$	Price increased by approximately 21.21% when compared with 1990
1993	$\dfrac{4400}{3300} \times 100 = 133.33$	Price increased by approximately 33.33% when compared with 1990
1994	$\dfrac{4600}{3300} \times 100 = 139.39$	Price increased by approximately 39.39% when compared with 1990
1995	$\dfrac{5000}{3300} \times 100 = 151.52$	Price increased by approximately 51.52% when compared with 1990

9. Pension costs in January 1994 were 6% less when compared with January 1990.

ANSWERS TO EXERCISES FOR SECTION 2.7

1. a) MTB > SET THE FOLLOWING DATA INTO C1
DATA > 430 441 460 420 437 450 426 460 438 456
DATA > 400 425 409 428 416 442 422 475 453 443
DATA > 450 410 422 406 404 460 429 420 446 405
DATA > 375 408 430 411 417 420 435 403 422 412
DATA > 493 430 401 429 406 409 442 427 409 459
DATA > END
MTB > GSTD
MTB > HISTOGRAM OF C1;
SUBC > START 381;
SUBC > INCREMENT 12.

Histogram of C1   N = 50

Midpoint	Count	
381.0	1	*
393.0	0	
405.0	12	* * * * * * * * * * * *
417.0	10	* * * * * * * * * *
429.0	9	* * * * * * * * *
441.0	8	* * * * * * * *
453.0	4	* * * *
465.0	4	* * * *
477.0	1	*
489.0	1	*

b) MTB > STEM AND LEAF OF C1

Stem-and-leaf of C1   N = 50
Leaf Unit = 1.0

1	37	5
1	38	
1	39	
12	40	01345668999
17	41	01267
(12)	42	000222567899
21	43	000578
15	44	12236
10	45	00369
5	46	000
2	47	5
1	48	
1	49	3

5. MTB > DOTPLOT OF C1

Character Dotplot

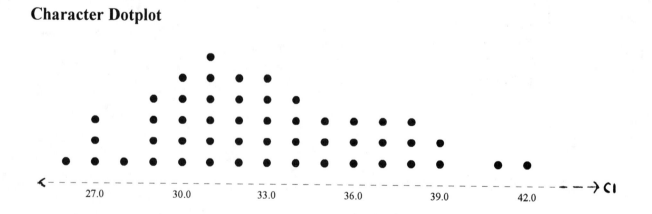

ANSWERS TO TESTING YOUR UNDERSTANDING OF THIS CHAPTER'S CONCEPTS - (PAGES 110 - 111)

1. Yes

ANSWERS TO CHAPTER TEST- (PAGES 111 - 118)

1. $0.14 \times 150000 = 21,000$ Choice (d)

5. 0 since its frequency is 0.   Choice (b)

9.

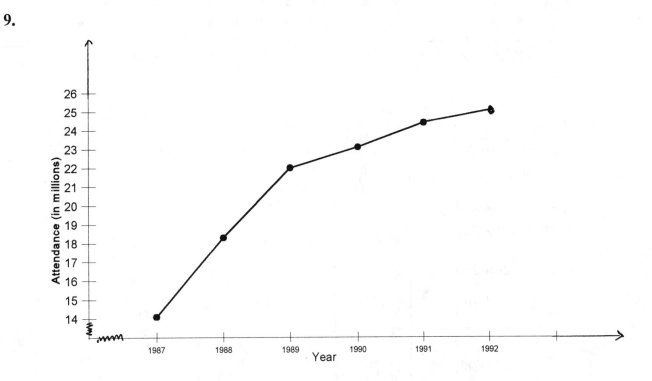

13.

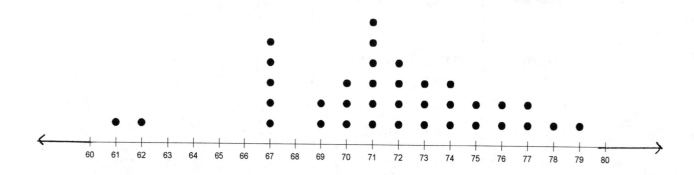

17. a)

Cholesterol Level	Tally	Frequency
224-234	⧕	5
235-245	⧕ ⧕ IIII	14
246-256	⧕ I	6
257-267	⧕ ⧕ IIII	14
268-278	IIII	4
279-289	⧕ III	8
290-300	II	2
301-311	II	2
312-322	III	3
323-333	II	<u>2</u>
		60

b. c)

Cholesterol Level	Percent Frequency Distribution	Cumulative Percent Frequency Distribution
224-234	$\dfrac{5}{60} = 8.33\%$	$\dfrac{5}{60} = 8.33\%$
235-245	$\dfrac{14}{60} = 23.33\%$	$\dfrac{19}{60} = 31.67\%$
246-256	$\dfrac{6}{60} = 10.00\%$	$\dfrac{25}{60} = 41.67\%$
257-267	$\dfrac{14}{60} = 23.33\%$	$\dfrac{39}{60} = 65.00\%$
268-278	$\dfrac{4}{60} = 6.67\%$	$\dfrac{43}{60} = 71.67\%$
279-289	$\dfrac{8}{60} = 13.33\%$	$\dfrac{51}{60} = 85.00\%$
290-300	$\dfrac{2}{60} = 3.33\%$	$\dfrac{53}{60} = 88.33\%$
301-311	$\dfrac{2}{60} = 3.33\%$	$\dfrac{55}{60} = 91.67\%$
312-322	$\dfrac{3}{60} = 5.00\%$	$\dfrac{58}{60} = 96.67\%$
323-333	$\dfrac{2}{60} = 3.33\%$	$\dfrac{60}{60} = 100.00\%$

d)

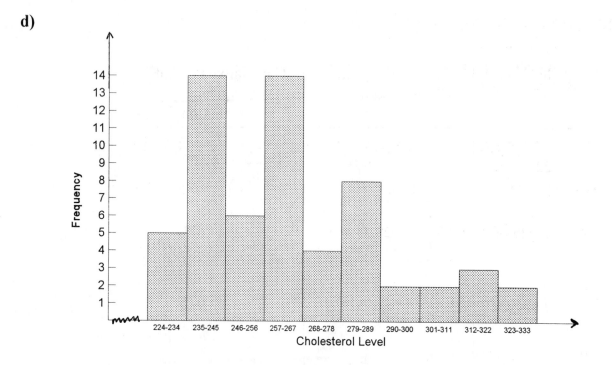

e)

Stem	Leaves
22	9 8 9 9 4
23	9 7 7 8 9 6
24	0 4 5 1 2 3 5 1 8 8
25	1 9 8 6 8 6 1
26	7 7 2 1 2 2 1 2 1 7 1
27	8 9 9 2 8 2
28	7 1 1 9 9 5
29	6 1
30	3 9
31	2 2
32	4 7 1
33	

21.   **a)** Normally distributed
     b) Probably not normally distributed
     c) Probably not normally distributed
     d) Normally distributed

ANSWERS TO THINKING CRITICALLY - (PAGES 118 - 120)

1. Yes. The next interval is supposed to begin where the previous interval ends. No

   gaps. Also, the interval lengths are not the same.

5.

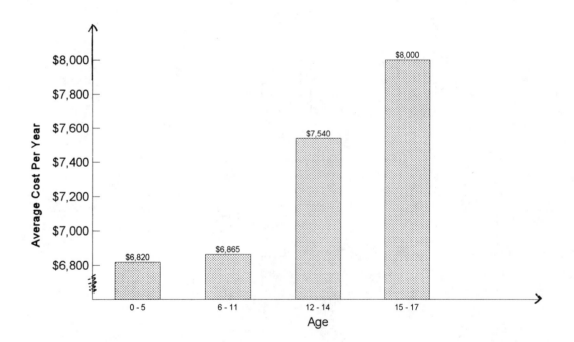

1. a)

Lead Levels	Tally	Frequency
13-14	|	1
15-16	||||	4
17-18	卌 卌 |	11
19-20	卌 卌 |	11
21-22	卌 卌 ||	12
23-24	卌 |||	8
25-26	||||	4
27-28	||||	4
29-30	|||	3
31-32	||	2
		60

b)

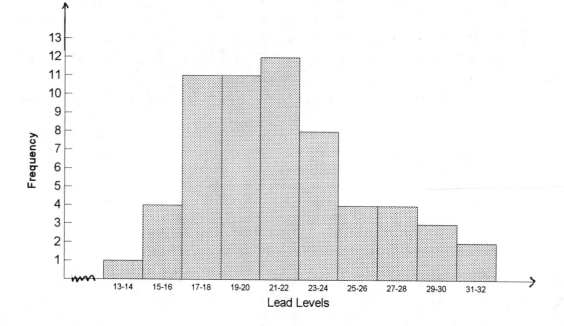

c)

Stem	Leaves
1	5 5 4
1	9 7 9 8 8 9 8 8 8 9 7 6 8 9 8 9 8 6 7
2	2 1 4 3 2 3 0 2 2 3 0 2 0 4 3 4 2 1 0 3 2 0 1 1 1
2	9 6 7 8 5 9 8 6 7 5
3	1 2 0

d)

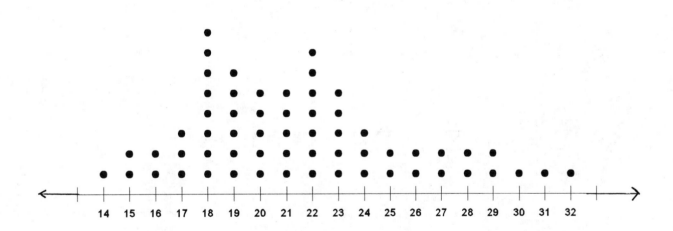

CHAPTER 3

ANSWERS TO EXERCISES FOR SECTION 3.2

1. a) $\displaystyle\sum_{i=1}^{7} x_i$

b) $\displaystyle\sum_{i=1}^{7} x_i^2$

c) $\displaystyle\sum_{i=1}^{7} x_i f_i$

d) $\displaystyle\sum_{i=1}^{7} 10x_i \quad$ or $\quad 10\sum_{i=1}^{7} x_i$

e) $\displaystyle\sum_{i=1}^{n} (2y_i + x_i)$

5. a) $\displaystyle\sum_{i=1}^{10} (x_i + 8) = \sum_{i=1}^{10} x_i + \sum_{i=1}^{10} 8 \; = 18 + 8 \cdot 10 = 98$

b)

$$\sum_{i=1}^{10} (x_i + 8)^2 = \sum_{i=1}^{10} (x_i^2 + 16x_i + 64)$$

$$= \sum_{i=1}^{10} x_i^2 + 16 \sum_{i=1}^{10} x_i + \sum_{i=1}^{10} 64 = 37 + 16(18) + 10 \cdot 64 = 965$$

9. a)

$$\sum x = 4.9 + 5.2 + 5.3 + 5.1 + 4.8 + 4.7 + 4.9 + 5.0 + 5.1 + 5.2 = 50.2$$

b) $(\sum x)^2 = (50.2)^2 = 2520.04$

c) $\sum x = 4.9^2 + 5.2^2 + 5.3^2 + 5.1^2 + 4.8^2 + 4.7^2 + 4.9^2 + 5.0^2 + 5.1^2 + 5.2^2 = 252.34$

ANSWERS TO EXERCISES FOR SECTION 3.3

1. Mean $= \dfrac{17 + 31 + 28 + 32 + 16 + 19 + 23 + 24 + 27 + 15}{10} = \dfrac{232}{10} = 23.2$ years

Median $= \dfrac{23 + 24}{2} = 23.5$

Mode $=$ None

5. Mean $= \dfrac{100,000 + 125,000 + \cdots + 130,000}{20} = \dfrac{2,558,000}{20} = \$127,900$

Median $= \dfrac{115,000 + 125,000}{2} = \$120,000$

Mode $=$ None

9. Average cost $= \dfrac{80 \times 60 + 100 \times 46 + 180 \times 48}{80 + 100 + 180} = \dfrac{18040}{360} = \$50.11$

13. $\dfrac{\$1,716,000}{22} = \$78,000$

17. Harmonic mean

$$= \dfrac{4}{\dfrac{1}{8} + \dfrac{1}{5} + \dfrac{1}{9} + \dfrac{1}{6}} = \dfrac{8640}{270 + 432 + 240 + 360} = \dfrac{8640}{1302} = 6.6359$$

Geometric mean $= \sqrt[4]{8 \cdot 5 \cdot 9 \cdot 6} = \sqrt[4]{2160} \approx 6.8173$

21.

$$\sum_{i=1}^{n} (x_i - \bar{x}) = \sum_{i=1}^{n} x_i - \sum_{i=1}^{n} \bar{x}$$

$$= \sum_{i=1}^{n} x_i - n\,\bar{x}$$

$$= \sum_{i=1}^{n} x_i - n\dfrac{\displaystyle\sum_{i=1}^{n} x_i}{n}$$

$$= \sum_{i+1}^{n} x_i - \sum_{i+1}^{n} x_i = 0$$

1.

x	$x - \mu$	$(x - \mu)^2$
14	$14 - 14 = 0$	$2^2 = 4$
11	$11 - 12 = -1$	$(-1)^2 = 1$
10	$10 - 12 = -2$	$(-2)^2 = 4$
17	$17 - 12 = 5$	$5^2 = 25$
12	$12 - 12 = 0$	$0^2 = 0$
8	$8 - 12 = -4$	$(-4)^2 = 16$
72	0	50

Mean $\mu = \dfrac{72}{6} = 12$

Range $= 17 - 8 = 9$

Population variance, $\sigma^2 = \dfrac{\sum (x - \mu)^2}{N} = \dfrac{50}{6} = 8.33333$

Population standard deviation, $\sigma = \sqrt{8.33333} \approx 2.88675$

5.

x	$x - \bar{x}$	$(x - \bar{x})^2$	$|x - \bar{x}|$
21	$21 - 25 = -4$	16	4
33	$33 - 25 = 8$	64	8
16	$16 - 25 = -9$	81	9
29	$29 - 25 = 4$	16	4
20	$20 - 25 = -5$	25	5
32	$32 - 25 = 7$	49	7
25	$25 - 25 = 0$	0	0
27	$27 - 25 = 2$	4	2
22	$22 - 25 = -3$	9	3
225	0	264	42

Sample mean $= \dfrac{225}{9} = 25$

Sample variance, $s^2 = \dfrac{264}{9-1} = 33$

Sample Standard deviation, $s = \sqrt{33} \approx 5.7446$

Average deviation $= \dfrac{42}{9} = 4.6667$

9.

x	x^2	$x - \bar{x}$	$|x - \bar{x}|$
16	256	16 - 20 = -4	4
8	64	8 - 20 = -12	12
32	1024	32 - 20 = 12	12
12	144	12 - 20 = -8	8
40	1600	40 - 20 = 20	20
36	1296	36 - 20 = 16	16
8	64	8 - 20 = -12	12
24	576	24 - 20 = 4	4
4	16	4 - 20 = -16	16
180	5040	0	104

New sample mean $= \dfrac{180}{9} = 20$

New sample variance $s^2 = \dfrac{9(5040) - 180^2}{9(9 - 1)} = \dfrac{12960}{72} = 180$

New sample standard deviation, $s = \sqrt{180} \approx 13.4164$

New average deviation $= \dfrac{104}{9} = 11.5556$

The sample variance is 16 times as great as it was originally.

The sample standard deviation and average deviation are 4 times as great as they were originally.

13. Probably Brand B.

1.

x	x^2
$584	341056
681	463761
500	250000
525	275625
586	343396
532	283024
3408	1,956,862

$$\bar{x} = \frac{\Sigma x}{n} = \frac{3408}{6} = 568$$

$$s = \sqrt{\frac{6(1,956,862) - (3408)^2}{6(6 - 1)}} = \sqrt{4223.6} \approx 64.989$$

When $k = 2$, then at least $\frac{3}{4}$ of the terms fall between $568 \pm 2(64.989)$ or between 438.022 and 697.978.

When $k = 3$, then at least $\frac{8}{9}$ of the terms fall between $568 \pm 3(64.989)$ or between 373.033 and 762.967.

5.

x	x^2
4.2	17.64
1.3	1.69
1.8	3.24
2.1	4.41
3.2	10.24
3.7	13.69
2.9	8.41
3.5	12.25
5.7	32.49
4.1	16.81
3.4	11.56
3.8	14.44
2.8	7.84
2.9	8.41
3.2	10.24
3.1	9.61
5.1	26.01
2.3	5.29
6.1	37.21
1.1	1.21
66.3	252.69

a) $\quad \bar{x} = \dfrac{66.3}{20} = 3.315$

$$s^2 = \frac{20(252.69) - (66.3)^2}{20(20 - 1)} = \frac{658.11}{380} \approx 1.7319$$

$$s = \sqrt{1.7319} \approx 1.3160$$

b) When $k = 1.75$, then at least $1 - \dfrac{1}{(1.75)^2} = 0.67$ or 67% of the terms fall within the interval $\bar{x} \pm 1.75s = 3.315 \pm 1.75(1.3160)$ or between 1.012 and 5.618.

c) All the measurements fall within the specified intervals.

d) Chebyshev's Theorem is valid.

ANSWERS TO EXERCISES FOR SECTION 3.7

1. Percentile rank of Michele $= \dfrac{8 + \frac{1}{2}(2)}{20} \cdot 100 = 45^{th}$ percentile.

Percentile rank of Ricardo $= \dfrac{17 + \frac{1}{2}(1)}{20} \cdot 100 = 87.5^{th}$ percentile.

5.

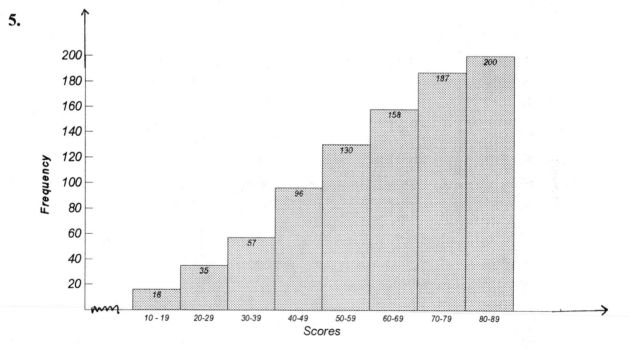

b)

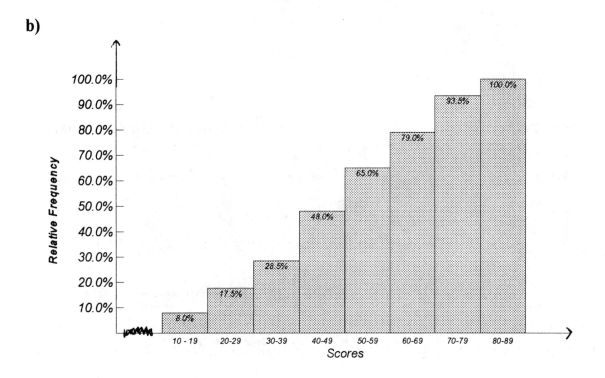

9. a)

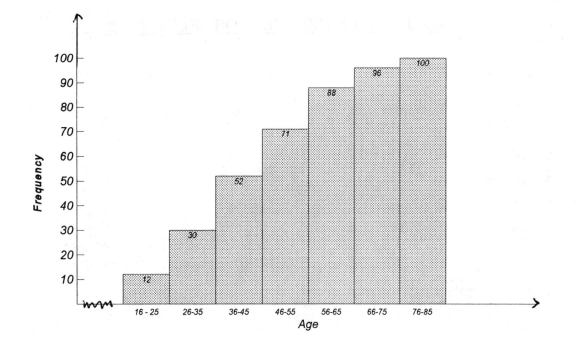

ANSWERS TO EXERCISES FOR SECTION 3.7

b)

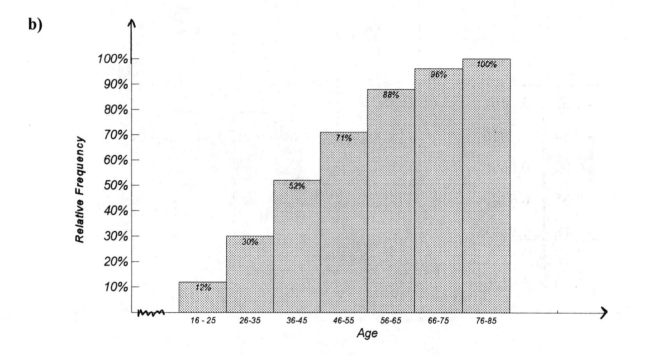

ANSWERS TO EXERCISES FOR SECTION 3.8

1.

a) $z = \dfrac{250 - 210}{20} = 2$

b) $z = \dfrac{180 - 210}{20} = -1.5$

c) $z = \dfrac{270 - 210}{20} = 3$

d) $z = \dfrac{210 - 210}{20} = 0$

5.

x	x^2
-1.52	2.3104
-0.54	0.2916
0.00	0.0000
0.65	0.4225
1.41	1.9881
0	5.0126

$$\sigma = \sqrt{\frac{5.0126}{5} - \frac{0^2}{25}} = \sqrt{1.00252} \approx 1$$

The slight difference from 1 is due to rounding.

x	x^2
-1.32	1.7424
-0.66	0.4356
0.00	0.0000
0.33	0.1089
1.65	2.7225
0.0	5.0094

$$\sigma = \sqrt{\frac{5.0094}{5} - \frac{0^2}{25}} = \sqrt{1.00188} \approx 1$$

The slight difference from 1 is due to rounding.

ANSWERS TO EXERCISES FOR SECTION 3.8

9.

New York Lake	Rating, x	x^2	z - score
A	22	484	-0.51
B	15	225	-1.49
C	29	841	0.47
D	33	1089	1.04
E	24	576	-0.23
F	17	289	-1.21
G	37	1369	1.60
H	28	784	0.33
	205	5657	

$$= \sqrt{\frac{5657}{8} - \frac{(205)^2}{64}} = \sqrt{50.48438}$$

$$\approx 7.11$$

$$= \frac{205}{8} = 25.625$$

New Jersey Lake	Rating, x	x^2	z - score
Q	68	4624	-0.405
R	75	5625	1.01
S	61	3721	-1.82
T	70	4900	0.000
U	70	4900	0.000
V	76	5776	1.216
	420	29546	

$$= \sqrt{\frac{29546}{6} - \frac{(420)^2}{36}} = \sqrt{24.3333}$$

$$\approx 4.933$$

$$= \frac{420}{6} = 70$$

ANSWERS TO EXERCISES FOR SECTION 3.8

a) New York Lake G since it has the highest z - value.

b) New Jersey Lake S since it has the smallest z - value

ANSWERS TO EXERCISES FOR SECTION 3.9

1. a)
```
MTB  > SET THE FOLLOWING IN C1
DATA > 29500  28700  25500  20500  23000  29750
DATA > 23000  22000  22000  26100  24000  21000
DATA > 28500  20500  21000  23000  21750  23000
DATA > 27400  22750  27500  22750  25000  28000
DATA > 26000  19700  26750  22000  26000  21500
DATA > 25500  27500  25000  24000
DATA > END
MTB  > DESCRIBE C1
```

Descriptive Statistics

Variable	N	Mean	Median	TrMean	StDev	SEMean
C1	34	24416	24000	24357	2859	490

Variable	Min	Max	Q1	Q3
C1	19700	29750	22000	26913

b)

```
MTB  > BOXPLOT OF C1
```

Character Boxplot

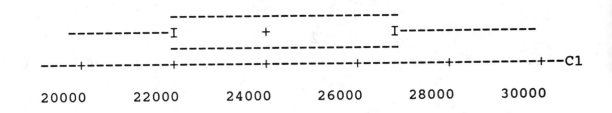

37

    c)   Interquatrile range = Q_3 - Q_1 = 26913 - 22000 = 4913

ANSWERS TO TESTING YOUR UNDERSTANDING OF THIS CHAPTER'S CONCEPTS - (PAGES 194 - 195)

1. b)
                c)

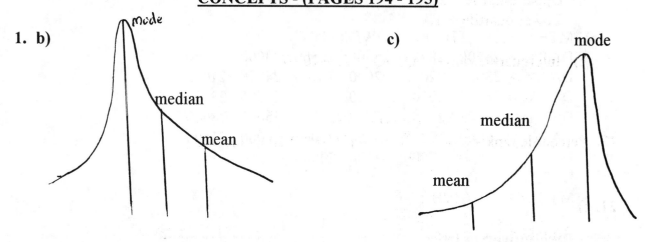

5. No.   Both schools do not necessarily have the same number of students.

9. 69° since this represents the mode.

ANSWERS TO CHAPTER TEST- (PAGES 195 - 200)

1. $\Sigma x^2 = 5^2 + 8^2 + 9^2 + 10^2 + 13^2 = 439$ Choice (b)

5. Heather's percentile rank = $\dfrac{2 + \frac{1}{2}(2)}{10} \cdot 100 = 30^{th}$ percentile.   Choice (d).

9. Standard deviation $= \sqrt{0.81} = 0.9$ Choice (a)

13. a) Median ≈ 22
  b) Upper quartile ≈ 23
    Lower quartile ≈ 16

  c) Interquartile range $= Q_3 - Q_1 = 23 - 16 = 7$

17. Percentile rank $= \dfrac{32 + \frac{1}{2}(5)}{88} \cdot 100 = 39.20$ percentile.

21. a)

Number of Times	Class mark, x	f	$x \cdot f$	$x^2 \cdot f$
0-1	0.5	125	12.5	31.25
2-3	2.5	85	212.5	531.25
4-5	4.5	65	292.5	1316.25
6-7	6.5	57	370.5	2408.25
8-9	8.5	50	425.0	3612.50
10-11	10.5	15	157.5	1653.75
12-13	12.5	14	175.0	2187.50
14-15	14.5	4	72.5	841.00
		415	1718	12581.75

Sample mean $= \dfrac{1718}{415} = 4.1398$

Sample standard deviation

$$= \sqrt{\frac{415(12581.75) - (1718)^2}{415(415 - 1)}} = \sqrt{13.2117} \approx 3.6348$$

b) $P_{55} = \dfrac{55}{100} \cdot 415 = 228.25$ Using cumulative frequencies, this falls in the 4-5

category. Using the class mark, $P_{55} = 4.5$

25. Probably the one from Company B since the standard deviation is smaller. Others may prefer the one from Company A since the average life is greater.

ANSWERS TO THINKING CRITICALLY- (PAGES 200 - 201)

1. Yes, if all the numbers are the same

5. Not necessarily true. See the data given in Exercise 4 of Section 3.7.

1.

x	x^2
$60.49	3659.0401
69.23	4792.7929
68.32	4667.6224
70.59	4982.9481
61.17	3741.7689
68.61	4707.3321
66.88	4472.9344
71.26	5077.9876
76.00	5776.0000
74.19	5504.1561
72.28	5224.3984
63.14	3986.6596
822.16	56593.6376

Sample mean $= \dfrac{822.16}{12} = \$68.51$

Sample standard deviation

$$= \sqrt{\dfrac{12(56593.6376) - (822.16)^2}{12(12 - 1)}}$$

$$= \sqrt{24.064993} \approx 4.9056$$

5.

Interval	Class mark, x	f	$x \cdot f$	$x^2 \cdot f$
0 - under 200	100	6	600	60,000
200 - under 400	300	7	2100	630,000
400 - under 600	500	16	8000	4,000,000
600 - under 800	700	32	22400	15,680,000
800 - under 1000	900	21	18900	17,010,000
1000 - under 1200	1100	15	16500	18,150,000
1200 - under 1400	1300	3	3900	5,070,000
		100	72400	60,600,000

$$\text{Sample mean} = \frac{72400}{100} = 724$$

$$\text{Sample variance} = \frac{100(60,600,000) - (72400)^2}{100(100-1)} = \frac{818,240,000}{9900}$$

$$= 82,650.505$$

$$\text{Sample standard deviation} = \sqrt{82,650.505} \approx 287.49$$

42

CHAPTER 4

ANSWERS TO EXERCISES FOR SECTION 4.2

1.

a) $\dfrac{137 + 102 + 98 + 81}{1032} = \dfrac{418}{1032} = \dfrac{209}{516}$

b) $\dfrac{19 + 49 + 137 + 157}{1032} = \dfrac{362}{1032} = \dfrac{181}{516}$

c) $\dfrac{17 + 16 + 12 + 28 + 4 + 1 + 102 + 98 + 81 + 206 + 59 + 46}{1032} = \dfrac{670}{1032} = \dfrac{335}{516}$

d) $\dfrac{19 + 17 + 16 + 12 + 49 + 28 + 4 + 1}{1032} = \dfrac{146}{1032} = \dfrac{73}{516}$

5.

a) $\dfrac{48 + 27 + 53}{300} = \dfrac{128}{300} = \dfrac{32}{75}$

b) $\dfrac{27 + 19 + 22}{300} = \dfrac{68}{300} = \dfrac{17}{75}$

c) $\dfrac{27}{300} = \dfrac{9}{100}$

d) $\dfrac{0}{300} = 0$

9.

a) $\dfrac{7 + 8}{38 + 47} = \dfrac{15}{85} = \dfrac{3}{17}$

b) $\dfrac{38}{85}$

c) $\dfrac{7}{85}$

13.

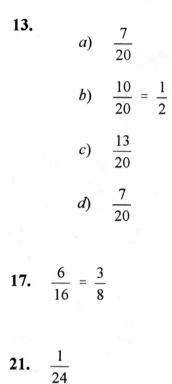

a) $\dfrac{7}{20}$

b) $\dfrac{10}{20} = \dfrac{1}{2}$

c) $\dfrac{13}{20}$

d) $\dfrac{7}{20}$

17. $\dfrac{6}{16} = \dfrac{3}{8}$

21. $\dfrac{1}{24}$

ANSWERS TO EXERCISES FOR SECTION 4.3

1. $3 \times 7 \times 2 = 42$ possible set-ups.

5. $52 \times 51 \times 50 = 132{,}600$ possible ways

9. $8 \times 10 \times 10 \times 10 \times 10 \times 10 \times 10 = 8{,}000{,}000$ possible numbers

13. a) $3 \times 4 \times 4 \times 4 = 192$ possible numbers

b) $3 \times 3 \times 2 \times 1 = 18$ possible numbers

17.

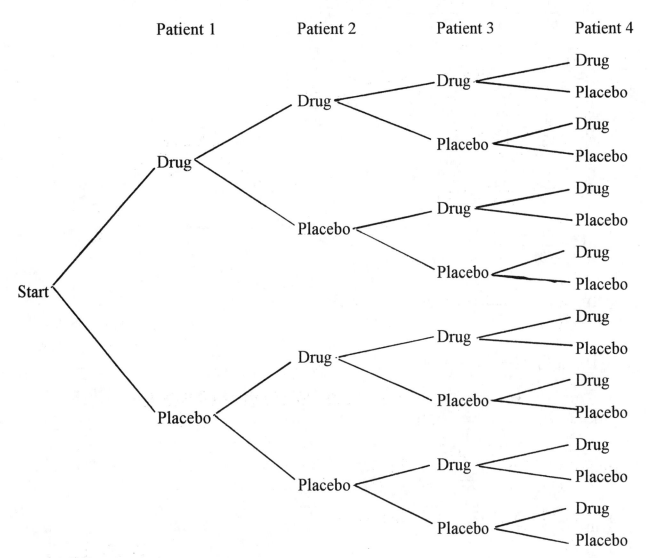

16 different ways

ANSWERS TO EXERCISES FOR SECTION 4.4

1. a) $6! = 6 \cdot 5 \cdot 4 \cdot 3 \cdot 2 \cdot 1 = 720$

b) $7! = 7 \cdot 6 \cdot 5 \cdot 4 \cdot 3 \cdot 2 \cdot 1 = 5040$

c) $2! = 2 \cdot 1 = 2$

d) $\dfrac{7!}{6!} = \dfrac{7 \cdot 6 \cdot 5 \cdot 4 \cdot 3 \cdot 2 \cdot 1}{6 \cdot 5 \cdot 4 \cdot 3 \cdot 2 \cdot 1} = 7$

e) $\dfrac{0!}{3} = \dfrac{1}{3}$

f) $\dfrac{6!}{4!\,2!} = 15$

g) $\dfrac{8!}{6!\,2!} = 28$

h) $\dfrac{6!}{3!\,3!} = 20$

i) ${}_7P_5 = \dfrac{7!}{(7-5)!} = 2520$

j) ${}_6P_4 = \dfrac{6!}{(6-4)!} = 360$

k) ${}_7P_3 = \dfrac{7!}{(7-3)!} = 210$

l) ${}_5P_4 = \dfrac{5!}{(5-4)!} = 120$

m) ${}_6P_6 = 6! = 720$

n) ${}_4P_4 = 4! = 24$

o) ${}_0P_0 = 0! = 1$

p) ${}_5P_0 = \dfrac{5!}{(5-0)!} = 1$

5. ${}_7P_3 = \dfrac{7!}{(7-3)!} = \dfrac{7!}{4!} = 210$ possible ways assuming order counts

ANSWERS TO EXERCISES FOR SECTION 4.4

9. a) $_{12}P_{12} = 12! = 479,001,600$

b) The men can be seated in $_6P_6 = 720$ possible ways and the women can also be seated in $_6P_6$ or 720 possible ways. However, the men can be seated on either the left side or right side, so that there are $720 \times 720 \times 2 = 1,036,800$ possible ways.

c) The couples can be seated in $_6P_6 = 6!$ Ways. However each man can sit on the left or right side of his wife. Thus, there are $2^6 \times 720 = 46,080$ possible ways.

13. The letters T, R, and O are each repeated twice. If the first letter is an O, then there are $\dfrac{11!}{2! \ 2!}$ or 9,979,200 possible ways. However, either of the O's can be the first letter we have $2 \times 9,979,200 = 19,958,400$ possible ways. If the first letter is an I, there are $\dfrac{11!}{2! \ 2! \ 2!} = 4,989,600$ possible ways. If the first letter is an E, there are

$\dfrac{11!}{2! \ 2! \ 2!} = 4,989,600$ possible ways. Thus, there are

$19,958,400 + 4,989,600 + 4,989,600 = 29,937,600$ possible ways.

17. $_7P_7 = 5040$ possible ways

1. a) $\quad _7C_6 = \dfrac{7!}{6!(7-6)!} = 7$

b) $\quad _6C_4 = \dfrac{6!}{4!\,2!} = 15$

c) $\quad _8C_3 = \dfrac{8!}{3!\,5!} = 56$

d) $\quad _9C_0 = \dfrac{9!}{0!\,9!} = 1$

e) $\quad _6C_1 = \dfrac{6!}{1!\,5!} = 6$

f) $\quad _8C_5 = \dfrac{8!}{5!\,3!} = 56$

g) $\quad \dbinom{9}{6} = \dfrac{9!}{6!\,3!} = 84$

h) $\quad \dbinom{8}{6} = \dfrac{8!}{6!\,2!} = 28$

i) $\quad \dbinom{7}{8} = \;_7C_8 = $ Impossible

j) $\quad \dbinom{6}{6} = \dfrac{6!}{6!\,0!} = 1$

5. $\quad _{10}C_4 = \dfrac{10!}{4!\,6!} = 210$

9. *a)* $\dfrac{_{13}C_2 \cdot _{11}C_0}{_{24}C_2} = \dfrac{78 \cdot 1}{276} = \dfrac{78}{276} = \dfrac{13}{46}$

b) $\dfrac{_{13}C_0 \cdot _{11}C_2}{_{24}C_2} = \dfrac{1 \cdot 55}{276} = \dfrac{55}{276}$

c) $\dfrac{_{13}C_1 \cdot _{11}C_1}{_{24}C_2} = \dfrac{13 \cdot 11}{276} = \dfrac{143}{276}$

13. a) $_{18}C_3 \cdot _2C_0 = 816 \cdot 1 = 816$

b) $_{18}C_2 \cdot _2C_1 = 153 \cdot 2 = 306$

c) $_{18}C_1 \cdot _2C_2 = 18 \cdot 1 = 18$

17 . a) $\dfrac{_{36}C_0 \cdot _{14}C_4}{_{50}C_4} = \dfrac{1 \cdot 1001}{230,300} = \dfrac{1001}{230,300}$

b) Committee can consist of

2 Republicans		3 Republicans		4 Republicans
2 Democrats	+	1 Democrat	+	0 Democrats

$$\dfrac{_{36}C_2 \cdot _{14}C_2}{_{50}C_4} + \dfrac{_{36}C_1 \cdot _{14}C_3}{_{50}C_4} + \dfrac{_{36}C_0 \cdot _{14}C_4}{_{50}C_4}$$

$$\dfrac{630 \cdot 91}{230,300} + \dfrac{36 \cdot 364}{230,300} + \dfrac{1 \cdot 1001}{230,300} =$$

$$\dfrac{57330}{230,300} + \dfrac{13104}{230,300} + \dfrac{1001}{230,300} = \dfrac{71435}{230,300} = \dfrac{2041}{6580}$$

ANSWERS TO EXERCISES FOR SECTION 4.5

c) $\dfrac{_{36}C_4 \cdot {}_{14}C_0}{_{50}C_4} = \dfrac{58905 \cdot 1}{230,300} = \dfrac{58,905}{230,300} = \dfrac{11,781}{46,060}$

ANSWERS TO EXERCISES FOR SECTION 4.6

1. $(+76)\left(\dfrac{5}{11}\right) + (-31)\left(\dfrac{6}{11}\right) = \dfrac{194}{11} = \$17.64$

5. $(+15)\left(\dfrac{12}{52}\right) + (+20)\left(\dfrac{4}{52}\right) + (-10)\left(\dfrac{36}{52}\right) = \dfrac{-100}{52} = -\$1.92$

9. **a)** 0.88

    b) 88:12 or 22:3

13. 79:21

ANSWERS TO EXERCISES FOR SECTION 4.7

1. **a)**   MTB > RANDOM 180 C1;
       SUBC > INTEGER 1 TO 6.
       MTB > PRINT C1
       MTB > TALLY C1
       Answers will vary

    b)   MTB > RANDOM 300 C1;
       SUBC >INTEGER 1 TO 6.
       MTB > PRINT C1
       MTB > TALLY C1
       Answers will vary

ANSWERS TO EXERCISES FOR SECTION 4.7

c)   MTB > RANDOM 600 C1;
SUBC > INTEGER 1 TO 6.
MTB > PRINT C1
MTB > TALLY C1
Answers will vary

d)   MTB > RANDOM 6000 C1;
SUBC > INTEGER 1 TO 6.
MTB > PRINT C1
MTB > TALLY C1
Answers will vary

e)   Yes

ANSWERS TO TESTING YOUR UNDERSTANDING OF THIS CHAPTER'S CONCEPTS- (PAGES 266- 267)

1. $\dfrac{1}{120}$

5. $10 \cdot 10 \cdot 10 \cdot 10 \cdot 10 \cdot 10 \cdot 10 \cdot 10 \cdot 10 = 1{,}000{,}000{,}000$ possible numbers. If a tenth

digit is added, there will be $10 \cdot 10 \cdot 10 \cdot 10 \cdot 10 \cdot 10 \cdot 10 \cdot 10 \cdot 10 \cdot 10$

$= 10{,}000{,}000{,}000$ possible numbers.

ANSWERS TO CHAPTER TEST - (PAGES 267 - 272)

1. $\dfrac{6}{36}$ Choice (b)

5.

Student 1	Student 2	Student 3		
Paper 1	Paper 2	Paper 3		
Paper 1	Paper 3	Paper 2	$\frac{1}{6}$	Choice (a)
Paper 2	Paper 1	Paper 3		
Paper 2	Paper 3	Paper 1		
Paper 3	Paper 1	Paper 2		
Paper 3	Paper 2	Paper 1		

9. $\frac{11}{100}$ Choice (a)

13. $30 \cdot 30 \cdot 30 = 27,000$

17. $8 \cdot 10 = 80$

21. 20 to 80 or 1 : 4

25. $5 \cdot 4 \cdot 3 = 60$

29. $\frac{3}{5}$

33. 85 : 15 or 17 : 3

37. a) 3 : 3 or 1 : 1

b) $(+7)\left(\frac{2}{6}\right) + (+8)\left(\frac{1}{6}\right) + (-3)\left(\frac{2}{6}\right) = \frac{16}{6} = \$2.67$

ANSWERS TO THINKING CRITICALLY- (PAGES 272 - 274)

1. No

9. No, the events do not have the same probability.

ANSWERS TO THE CASE STUDIES- (PAGES 274 - 275)

1. **a)** There are 25×999 or $24,975$ possible combinations. The probability of winning $= \dfrac{1}{24,975}$ assuming there are no duplications and that only one car will be awarded.

b) $24,975$ to 1

CHAPTER 5

ANSWERS TO EXERCISES FOR SECTION 5.2

1. **a)** Not mutually exclusive
    b) Mutually exclusive
    c) Mutually exclusive (assuming that a person buys only one ticket)
    d) Not mutually exclusive
    e) Mutually exclusive
    f) Mutually exclusive
    g) Not mutually exclusive
    h) Mutually exclusive

5. p(either is a college graduate) = 0.35 + 0.41 - 0.14 = 0.62

9. p(receives bachelor's or master's degree) = p(receives bachelor's degree) + p(receives master's degree) - p(receives both degrees)
$$0.96 = 0.70 + p(\text{receives master's degree}) - 0$$
$$0.96 - 0.70 = 0.26 = p(\text{receives master's degree})$$

13. p(jogs, swims, or cycles) = 0.53 + 0.44 + 0.46 - 0.18 - 0.15 - 0.17 + 0.07 = 1.00

ANSWERS TO EXERCISES FOR SECTION 5.3

1. *a)* $\dfrac{26 + 61 + 89 + 59 + 46}{433} = \dfrac{281}{433}$

    b) $\dfrac{59}{281}$

    c) $\dfrac{59}{59 + 37} = \dfrac{59}{96}$

5. p(provides laptop computer | provides desktop computer) = $\dfrac{0.23}{0.77} = \dfrac{23}{77}$

ANSWERS TO EXERCISES FOR SECTION 5.3

9. p(passes credit-bearing Eng. Course | passed competency exam)

$$= \frac{p(\text{passes both})}{p(\text{passes competency exam})}$$

$$0.88 = \frac{p(\text{passes both})}{0.75}$$

$$p(\text{passes both}) = (0.88)(0.75) = 0.66$$

ANSWERS TO EXERCISES FOR SECTION 5.4

1. p(driver under 30 years of age and car driven is uninsured) $= (0.39)(0.13) = 0.0507$

5. p(disk manufactured by Stanton Corp. and functions properly)
$= (0.93)(0.71) = 0.6603$

9. p(stopped by neither) $= p$(not stopped by guard) $\cdot$ p(not stopped by metal detector)
$= (0.26)(0.01) = 0.0026$

13. p(male not part-time and female not part-time) $= (1 - 0.46)(1 - 0.53) = 0.2538$

ANSWERS TO EXERCISES FOR SECTION 5.5

1. p(assembled by workers in Factory 3 | defective)

$$= \frac{(0.06)(0.10)}{(0.07)(0.60) + (0.03)(0.30) + (0.06)(0.10)} = \frac{0.006}{0.057} = \frac{2}{19}$$

ANSWERS TO EXERCISES FOR SECTION 5.5

5. $p(60 \text{ years or older} \mid \text{male}) = \dfrac{(0.40)(0.40)}{(0.40)(0.40) + (0.60)(0.42)} = \dfrac{0.160}{0.412} = \dfrac{40}{103}$

9. $p(\text{produced by company A or B} \mid \text{needs adjustment})$

$= \dfrac{(0.001)(0.10) + (0.005)(0.20)}{(0.001)(0.10) + (0.005)(0.20) + (0.001)(0.40) + (0.002)(0.30)} = \dfrac{0.0011}{0.0029} = \dfrac{11}{29}$

ANSWERS TO EXERCISES FOR SECTION 5.6

1. MTB < READ C1   C2
   DATA < 1   0.35
   DATA < 2   0.25
   DATA < 3   0.22
   DATA < 4   0.18
   DATA < END
             4 ROWS READ
   MTB   < RANDOM 100 VALUES AND PLACE IN C3;
   SUBC < DISCRETE SAMPLE VALUES FROM C1 WITH PROBABILITIES IN C2.
   MTB   < HISTOGRAM OF C3;
   SUBC < START   1;
   SUBC < INCREMENT 1.
   HISTOGRAM OF C3   N = 100

In this case, 1 represents Math, 2 represents Physics, 3 represents Chemistry and 4

represents Biology. The actual histogram will vary as each computer is generating

numbers randomly.

ANSWERS TO TESTING YOUR UNDERSTANDING OF THIS CHAPTER'S CONCEPTS - (PAGES 321 - 322)

1. Mutually exclusive events are two events that cannot occur at the same time.

   Independent events are two events where the likelihood of the occurrence of one

   event is in no way affected by the occurrence or nonoccurrence of the other event.

5. No

ANSWERS TO CHAPTER TEST- (PAGES 322 - 326)

1. $\dfrac{47 + 45 + 22 + 20}{270} = \dfrac{134}{270}$ Choice (b)

5. $p(\text{picture card or red card}) = p(\text{picture card}) + p(\text{ red card}) - p(\text{ red picture card})$

   $= \dfrac{12}{52} + \dfrac{26}{52} - \dfrac{6}{52} = \dfrac{32}{52}$ Choice (d)

9. $p(\text{student 1 is math major and student 2 is art major}) = (0.72)(0.51) = 0.3672$

13. $p(\text{both fish are catfish}) = \left(\dfrac{4}{17}\right)\left(\dfrac{3}{16}\right) = \dfrac{12}{272} = \dfrac{3}{68}$

17. $p(\text{will get a ticket} \mid \text{meter out of order}) = \dfrac{2/13}{4/9} = \dfrac{2}{13} \div \dfrac{4}{9} = \dfrac{18}{52} = \dfrac{9}{26}$

21. $p(\text{high cholesterol or overweight}) = 0.42 + 0.30 - 0.15 = 0.57$

25. a) p(person has disease | positive) =

$$\frac{(0.97)(0.04)}{(0.97)(0.04) + (0.08)(0.96)} = \frac{0.0388}{0.1156} = \frac{388}{1156} = \frac{97}{289}$$

b) p(does not have disease | negative) =

$$\frac{(0.92)(0.96)}{(0.92)(0.96) + (0.03)(0.04)} = \frac{0.8832}{0.8844} = \frac{2208}{2211}$$

ANSWERS TO THINKING CRITICALLY - (PAGES 326 - 327)

1. p(second different from first) = $\dfrac{364}{365}$, assuming no leap year.

5. $\dfrac{1}{5}$

CHAPTER 6

ANSWERS TO EXERCISES FOR SECTION 6.2

1. a) 0, 1, 2, ...

b) 0, 1, 2, ...

c) 0, 1, 2, ...

d) Any non-negative real number

e) Any real number

f) 0, 1, 2, ...

g) 0, 1, 2, ...(It actually depends upon the Arab country.)

5. 0.20   since all the probabilities must sum to 1.

9.

Number of drivers checking oil, x	$p(x)$
0	$(0.94)(0.94)(0.94)\ = 0.8306$
1	$3(0.06)(0.94)(0.94) = 0.1590$
2	$3(0.06)(0.06)(0.94) = 0.0102$
3	$(0.06)(0.06)(0.06)\ = \underline{0.0002}$
	1.0000

13. a)

Number of boys in family, x	$p(x)$
0	1/16
1	4/16
2	6/16
3	4/16
4	<u>1/16</u>
	16/16 = 1

b) $p(x$ has a value of at least 1$) = \dfrac{4}{16} + \dfrac{6}{16} + \dfrac{4}{9} + \dfrac{1}{16} = \dfrac{15}{16}$

17. a) $p(x = 4) = 0.38$

b) $p(x = 7) = 0$

c) $p(x \leq 3) = 0.32 + 0.04 = 0.36$

d) $p(x \leq 6) = 0.32 + 0.04 + 0.38 + 0.17 + 0.09 = 1$

e) $p(x \leq 4 \text{ or } x > 5) = 0.32 + 0.04 + 0.38 + 0.09 = 0.83$

ANSWERS TO EXERCISES FOR SECTION 6.4

1.

x	$p(x)$	$x \cdot p(x)$	x^2	$x^2 \cdot p(x)$	
10	0.06	0.6000	100	100(0.06)	= 6.000
20	0.11	2.2000	400	400(0.11)	= 44.000
30	0.26	7.8000	900	900(0.26)	= 234.000
40	0.21	8.4000	1600	1600(0.21)	= 336.000
50	0.19	9.5000	2500	2500(0.19)	= 475.000
60	0.17	10.2000	3600	3600(0.17)	= 612.000
		38.7			1707

Mean = $\mu = \Sigma x \cdot p(x) = 38.7$

Variance = $\sigma^2 = \Sigma x^2 \cdot p(x) - [\Sigma x \cdot p(x)]^2$

$\quad = 1707 - (38.7)^2 = 209.31$

Standard deviation = $\sigma = \sqrt{209.31} \approx 14.4675$

5.

x	$p(x)$	$x \cdot p(x)$	x^2	$x^2 \cdot p(x)$
200	0.04	8	40000	1600
300	0.06	18	90000	5400
400	0.11	44	160000	17600
500	0.16	80	250000	40000
600	0.18	108	360000	64800
700	0.21	147	490000	102900
800	0.24	192	640000	153600
		597		385,900

61

Mean = μ = $\Sigma x \cdot p(x)$ = 597

Variance = σ^2 = $\Sigma x^2 \cdot p(x) - [\Sigma x \cdot p(x)]^2$

$\qquad$ = 385,000 $-$ (597)^2 = 29491

Standard deviation = σ = $\sqrt{29491}$ $\approx$ 171.7294

9.

x	$p(x)$	$x \cdot p(x)$	x^2	$x^2 \cdot p(x)$
2	0.10	0.20	4	0.40
3	0.23	0.69	9	2.07
4	0.22	0.88	16	3.52
5	0.19	0.95	25	4.75
6	0.14	0.84	36	5.04
7	0.08	0.56	49	3.92
8	0.03	0.24	64	1.92
9	0.01	0.09	81	0.81
		4.45		22.43

Mean = μ = $\Sigma x \cdot p(x)$ = 4.45

Variance = σ^2 = $\Sigma x^2 \cdot p(x) - [\Sigma x \cdot p(x)]^2$

$\qquad$ = 22.43 $-$ (4.45)^2 = 2.6275

Standard deviation = σ = $\sqrt{2.6275}$ $\approx$ 1.6210

13. a) Expected income = (50000)(0.49) + (-23,000)(0.51) = $12,770

$\quad$ **b)** Yes. The expected income minus the cost for rain insurance is still a positive number.

1. p(exactly 3 on outpatient basis) $= \dfrac{6!}{3!\,3!}\,(0.60)^3(0.40)^3 = 0.2765$

ANSWERS TO EXERCISES FOR SECTION 6.5

5. a) p(view exactly 3 hours of TV commercials $= \dfrac{7!}{3!\,4!}\,(0.50)^3(0.50)^4 = 0.2734$

   b) p(at most 3 of them view 3 hours of TV commercials)

   $= \dfrac{7!}{0!\,7!}\,(0.50)^0(0.50)^7 + \dfrac{7!}{1!\,6!}\,(0.50)^1(0.50)^6 + \dfrac{7!}{2!\,5!}\,(0.50)^2(0.50)^5 + \dfrac{7!}{3!\,4!}\,(0.50)^3(0.50)^4$

   $= 0.0078 + 0.0547 + 0.1641 + 0.2734 = 0.5000$

   c) p(at least 3 of them view 3 hours of TV commercials)

   $= \dfrac{7!}{3!\,4!}\,(0.5)^3(0.5)^4 + \dfrac{7!}{4!\,3!}\,(0.5)^4(0.5)^3 + \dfrac{7!}{5!\,2!}\,(0.5)^5(0.5)^2 + \dfrac{7!}{6!\,1!}\,(0.5)^6(0.5)^1$

   $+ \dfrac{7!}{7!\,0!}\,(0.5)^7(0.5)^0 = 0.2734 + 0.2734 + 0.1641 + 0.0547 + 0.0078 = 0.7734$

9. a) p(all of them see gynecologist) $= \dfrac{9!}{9!\,0!} = (0.70)^9(0.30)^0 = 0.0404$

   b) p(4 of them see gynecologist) $= \dfrac{9!}{4!\,5!} = (0.70)^4(0.30)^5 = 0.0735$

13. p(six of them are overweight) $= \dfrac{8!}{6!\,2!}\,(0.63)^6(0.37)^2 = 0.2397$

ANSWERS TO EXERCISES FOR SECTION 6.6 - (PAGES 49 - 52)

1.

$\mu = 3000(0.09) = 270$

$\sigma = \sqrt{3000(0.09)(0.91)} \approx 15.6748$

5.

$\mu = 4730(0.38) = 1797.40$

$\sigma = \sqrt{4730(0.38)(0.62)} \approx 33.3825$

9.

$\mu = 1800(0.88) = 1584$

$\sigma = \sqrt{1800(0.88)(0.12)} \approx 13.7870$

ANSWERS TO EXERCISES FOR SECTION 6.7

1. $p(\text{at most 2 fires}) = p(0) = p(1) + p(2)$

$$= \frac{e^{-4}(4^0)}{0!} + \frac{e^{-4}(4^1)}{1!} + \frac{e^{-4}(4^2)}{2!}$$

$$= 0.0183 + 0.0733 + 0.1465 = 0.2381$$

5. $p(\text{at least 2 suicides}) = 1 - p(x = 0) - p(x = 1)$

$$= 1 - \frac{e^{1.5}(1.5)^0}{0!} - \frac{e^{1.5}(1.5)^1}{1!}$$

$$= 1 - 0.2231 - 0.3347 = 0.4422$$

9. $p(\text{at most 3 will call in sick}) = p(x = 0) + p(x = 1) + p(x = 2) + p(x = 3)$

$$= \frac{e^{-4.8}(4.8)^0}{0!} + \frac{e^{-4.8}(4.8)^1}{1!} + \frac{e^{-4.8}(4.8)^2}{2!} + \frac{e^{-4.8}(4.8)^3}{3!}$$

$$= 0.0082 + 0.0395 + 0.0948 + 0.1517 = 0.2942$$

1. ***a)*** $\dfrac{\dbinom{15}{2}\dbinom{85}{8}}{\dbinom{100}{10}} = 0.2919$

b) $\dfrac{\dbinom{15}{0}\dbinom{85}{10} + \dbinom{15}{1}\dbinom{85}{9} + \dbinom{15}{2}\dbinom{85}{8}}{\dbinom{100}{10}} = 0.8295$

5. $\dfrac{\dbinom{15}{2}\dbinom{10}{3}}{\dbinom{25}{5}} = \dfrac{(105)(120)}{53130} = \dfrac{12600}{53130} = \dfrac{1260}{5313}$

9. $\dfrac{\dbinom{470}{25}\dbinom{30}{5}}{\dbinom{500}{30}} = 0.0211$

ANSWERS TO EXERCISES FOR SECTION 6.9

1.   MTB   >   PDF;
       SUBC   >   BINOMIAL   WITH N = 10   p = 0.60.

        BINOMIAL WITH $n = 10$, $p = 0.600000$

x	$P(X = x)$
0	0.0001
1	0.0016
2	0.0106
3	0.0425
4	0.1115
5	0.2007
6	0.2508
7	0.2150
8	0.1209
9	0.0403
10	0.0060

5.   MTB > PDF;
   SUBC > POISSON 11.4.

POISSON WITH MU = 11.4000

x	$P(X = x)$
0	0.0000
1	0.0001
2	0.0007
3	0.0028
4	0.0079
5	0.0180
6	0.0341
7	0.0556
8	0.0792
9	0.1003
10	0.1144
11	0.1185
12	0.1126
13	0.0987
14	0.0804
15	0.0611
16	0.0435
17	0.0292
18	0.0185
19	0.0111
20	0.0063
21	0.0034
22	0.0018
23	0.0009
24	0.0004
25	0.0002
26	0.0001
27	0.0000

1. a)

x	$p(x)$	$x \cdot p(x)$	x^2	$x^2 \cdot p(x)$
4	4/36	16/36	16	64/36
5	4/36	20/36	25	100/36
6	5/36	30/36	36	180/36
7	6/36	42/36	49	294/36
8	7/36	56/36	64	448/36
9	4/36	36/36	81	324/36
10	3/36	30/36	100	300/36
11	2/36	22/36	121	242/36
12	1/36	12/36	144	144/36
		264/36		2096/36

b)

$$\mu = \text{Mean} = \frac{264}{36} = \frac{22}{3}$$

$$\sigma^2 = \text{variance} = \frac{2096}{36} - \left(\frac{22}{3}\right)^2 = 4.4444$$

$$\sigma = \text{Standard deviation} = \sqrt{4.4444} \approx 2.108$$

68

5.

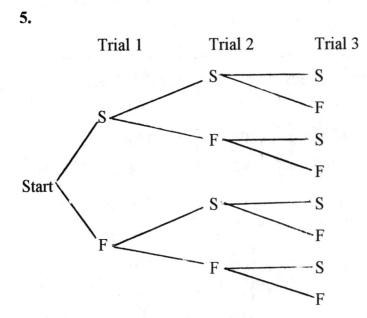

Trial 1   Trial 2   Trial 3

Where S = Success and F = Failure

<u>**ANSWERS TO CHAPTER TEST - (PAGES 392 - 396)**</u>

1. $1 - (0.18 - 0.32 - 0.25 - 0.17) = 0.08$ Choice (d)

5.

$\mu = \Sigma x \cdot p(x) = 5(0.07) + 6(0.14) + 7(0.17) + 8(0.21) + 9(0.18) + 10(0.19) + 11(0.04) = 8.02$

Choice (d)

9. $p(\text{at least six}) = p(\text{six}) + p(\text{seven}) + p(\text{eight}) + p(\text{nine}) + p(\text{ten})$

$$= \frac{10!}{6!\,4!}\left(\frac{1}{7}\right)^6\left(\frac{6}{7}\right)^4 + \frac{10!}{7!\,3!}\left(\frac{1}{7}\right)^7\left(\frac{6}{7}\right)^3 + \frac{10!}{8!\,2!}\left(\frac{1}{7}\right)^8\left(\frac{6}{7}\right)^2 + \frac{10!}{9!\,1!}\left(\frac{1}{7}\right)^9\left(\frac{6}{7}\right)^1$$

$$+ \frac{10!}{10!\,0!}\left(\frac{1}{7}\right)^{10}\left(\frac{6}{7}\right)^0 = 0.0010 + 0.0001 + 0.0000 + 0.0000 + 0.0000 = 0.0011$$

13. $p(\text{at most } 5) = p(x = 0) + p(x = 1) + p(x = 2) + p(x = 3) + p(x = 4) + p(x = 5)$

$$= \frac{e^{-7} 7^0}{0!} + \frac{e^{-7} 7^1}{1!} + \frac{e^{-7} 7^2}{2!} + \frac{e^{-7} 7^3}{3!} + \frac{e^{-7} 7^4}{4!} + \frac{e^{-7} 7^5}{5!}$$

$$= 0.0009 + 0.0064 + 0.0223 + 0.0521 + 0.0912 + 0.1277 = 0.3006$$

17. $p(\text{at most four}) = p(x = 0) + p(x = 1) + p(x = 2) + p(x = 3) + p(x = 4)$

$$= \frac{8!}{0!\,8!}(0.52)^0 (0.48)^8 + \frac{8!}{1!\,7!}(0.52)^1 (0.48)^7 + \frac{8!}{2!\,6!}(0.52)^2 (0.48)^6 + \frac{8!}{3!\,5!}(0.52)^3 (0.48)^5$$

$$+ \frac{8!}{4!\,4!}(0.52)^4 (0.48)^4 = 0.0028 + 0.0244 + 0.0926 + 0.2006 + 0.2717 = 0.5921$$

21. $p(\text{at most } 3) = p(x = 0) + p(x = 1) + p(x = 2) + p(x = 3)$

$$= \frac{8!}{0!\,8!}(0.20)^0 (0.80)^8 + \frac{8!}{1!\,7!}(0.20)^1 (0.80)^7 + \frac{8!}{2!\,6!}(0.20)^2 (0.80)^6 + \frac{8!}{3!\,5!}(0.20)^3 (0.80)^5$$

$$= 0.1678 + 0.3355 + 0.2936 + 0.1468 = 0.9437$$

25.

$$\mu = np = 80\left(\frac{1}{4}\right) = 20$$

$$\sigma = \sqrt{80\left(\frac{1}{4}\right)\left(\frac{3}{4}\right)} = \sqrt{15} \approx 3.8730$$

1.

x	$p(x)$
0	$\left(\dfrac{3}{8}\right)\left(\dfrac{3}{8}\right) = \dfrac{9}{64}$
1	$2\left(\dfrac{5}{8}\right)\left(\dfrac{3}{8}\right) = \dfrac{30}{64}$
2	$\left(\dfrac{5}{8}\right)\left(\dfrac{5}{8}\right) = \dfrac{25}{64}$

5.

$$\sigma = \Sigma(x - \mu)^2 \cdot p(x)$$

$$= \Sigma(x^2 - 2\mu x + \mu^2) \cdot p(x)$$

$$= \Sigma x^2 \cdot p(x) - 2\mu \Sigma x \cdot p(x) + \mu^2 \Sigma p(x) \quad \text{Note: } \mu = \Sigma x \cdot p(x) \text{ and } \Sigma p(x) = 1$$

$$= \Sigma x^2 \cdot p(x) - 2\mu\mu + \mu^2$$

$$= \Sigma x^2 \cdot p(x) - \mu^2$$

$$= \Sigma x^2 \cdot p(x) - [\Sigma x \cdot p(x)]^2$$

ANSWERS TO CASE STUDY - (PAGES 398 - 399)

1. $p(\text{at most 4 of them}) = p(x = 0) + p(x = 1) + p(x = 2) + p(x = 3) + p(x = 4)$

$$= \frac{15!}{0!\,15!}\left(\frac{1}{15}\right)^0\left(\frac{14}{15}\right)^{15} + \frac{15!}{1!\,14!}\left(\frac{1}{15}\right)^1\left(\frac{14}{15}\right)^{14} + \frac{15!}{2!\,13!}\left(\frac{1}{15}\right)^2\left(\frac{14}{15}\right)^{13}$$

$$+ \frac{15!}{3!\,12!}\left(\frac{1}{15}\right)^3\left(\frac{14}{15}\right)^{12} + \frac{15!}{4!\,11!}\left(\frac{1}{15}\right)^4\left(\frac{14}{15}\right)^{11}$$

$$= 0.3553 + 0.3806 + 0.1903 + 0.0589 + 0.0126 = 0.9977$$

CHAPTER 7

ANSWERS TO EXERCISES FOR SECTION 7.3

1. a) 0.4656

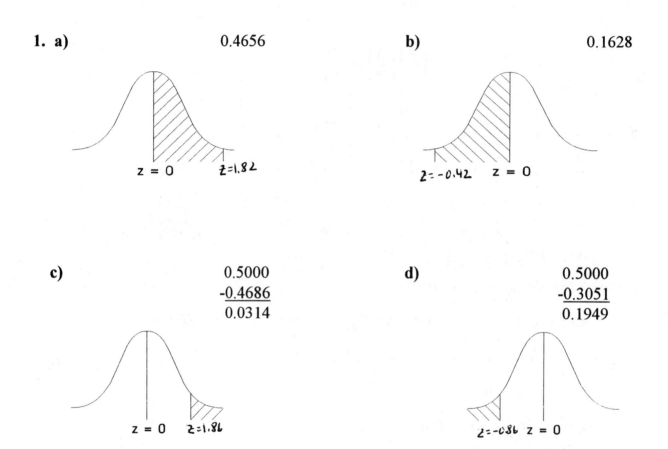

    z = 0   z = 1.82

b) 0.1628

    z = -0.42   z = 0

c)
$$0.5000$$
$$-0.4686$$
$$0.0314$$

    z = 0   z = 1.86

d)
$$0.5000$$
$$-0.3051$$
$$0.1949$$

    z = -0.86   z = 0

e)
$$0.5000$$
$$-0.4916$$
$$0.0084$$

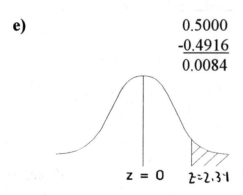

    z = 0   z = 2.34

f)
$$0.4495$$
$$0.4699$$
$$0.9194$$

    z = -1.64   z = 0   z = 1.88

g)
$$\begin{array}{r} 0.4625 \\ +0.4968 \\ \hline 0.9593 \end{array}$$

h)
$$\begin{array}{r} 0.4671 \\ -0.2019 \\ \hline 0.2652 \end{array}$$

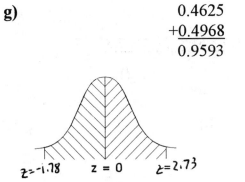

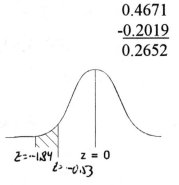

5. a)   z = 2.07

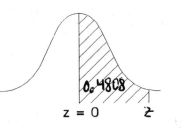

b)   z = 2.18

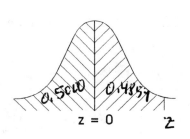

c)   z = -0.83

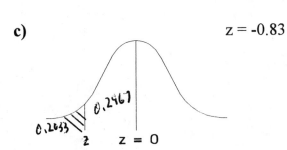

d)   z = 2.68

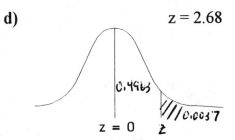

e)   z = 2.27

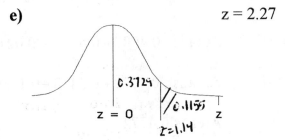

f)   z = 2.93

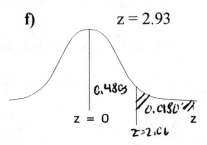

9.

$$z = 0.52$$
$$x = \mu + z\sigma$$
$$80 = 72 + (0.52)\sigma$$
$$\sigma = 15.3846$$

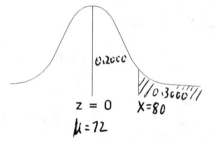

13.

$x = 18$ gives $z = \dfrac{18 - 27}{3} = -3$

$x = 21$ gives $z = \dfrac{21 - 27}{3} = -2$

$x = 24$ gives $z = \dfrac{24 - 27}{3} = -1$

$x = 27$ gives $z = \dfrac{27 - 27}{3} = 0$

$x = 30$ gives $z = \dfrac{30 - 27}{3} = 1$

$x = 33$ gives $z = \dfrac{33 - 27}{3} = 2$

$x = 36$ gives $z = \dfrac{36 - 27}{3} = 3$

The area of region I is between $z = -3$ and $z = -2$ Area is $0.4987 - 0.4772 = 0.0215$

Area of region II is between $z = -2$ and $z = -1$ Area is $0.4772 - 0.3413 = 0.1359$

Area of region III is between $z = -1$ and $z = 0$ Area is 0.4772

ANSWERS TO EXERCISES FOR SECTION 7.3

Area of region V is between $z = 1$ and $z = 2$ Area is $0.4772 - 0.3413 = 0.1359$

Area of region VI is between $z = 2$ and $z = 3$ Area is $0.4987 - 0.4772 = 0.0215$

ANSWERS TO EXERCISES FOR SECTION 7.4

1. $z = \dfrac{20 - 26}{4} = -1.5$

$$\begin{array}{r} 0.5000 \\ -0.4332 \\ \hline 0.0668 \end{array}$$

5. $z = \dfrac{9 - 8}{1.3} = 0.77$

$$\begin{array}{r} 0.5000 \\ -0.2794 \\ \hline 0.2206 \end{array}$$

9. $x = 5 + (-1.645)(1.1) = 3.1905$ years

13. $x = 180 + (1.04)(25)$ $x = 206$ minutes

17. $55 = 45 + (1.04)\,\sigma$ $\sigma = 9.6154$

ANSWERS TO EXERCISES FOR SECTION 7.5

1.

$\mu = 500(0.24) = 120$ $z = \dfrac{100.5 - 120}{9.5499} = -2.04$

$$\begin{array}{r} 0.5000 \\ -0.4793 \\ \hline 0.0207 \end{array}$$

$\sigma = \sqrt{500(0.24)(0.76)}$

≈ 9.5499

5.

$$\mu = 200(0.04) = 8 \qquad z = \frac{7.5 - 8}{2.7713} = -0.18 \qquad \begin{array}{r} 0.0714 \\ +0.5000 \\ \hline 0.5714 \end{array}$$

$$\sigma = \sqrt{200(0.04)(0.96)}$$

$$\approx 2.7713$$

9.

$$\mu = 80(0.35) = 28 \qquad z = \frac{20.5 - 28}{4.2661} = -1.76 \qquad \begin{array}{r} 0.4767 \\ -0.4608 \\ \hline 0.0159 \end{array}$$

$$\sigma = \sqrt{80(0.35)(0.65)} \qquad z = \frac{19.5 - 28}{4.2661} = -1.99$$

$$\approx 4.2661$$

13.

$$\mu = 480\left(\frac{2}{3}\right) = 320 \qquad z = \frac{299.5 - 320}{10.3280} = -1.98 \qquad \begin{array}{r} 0.5000 \\ -0.4761 \\ \hline 0.0239 \end{array}$$

$$\sigma = \sqrt{480\left(\frac{2}{3}\right)\left(\frac{1}{3}\right)}$$

$$\approx 10.3280$$

17.

$$\mu = 300\left(\frac{2}{5}\right) = 120 \qquad z = \frac{120.5 - 120}{8.4853} = 0.06 \qquad \begin{array}{r} 0.5000 \\ -0.0239 \\ \hline 0.4761 \end{array}$$

$$\sigma = \sqrt{300\left(\frac{2}{5}\right)\left(\frac{3}{5}\right)}$$

$$\approx 8.4853$$

ANSWERS TO EXERCISES FOR SECTION 7.7

1. **a)** 0.022750

   b) 1.000000 - 0.841345 = 0.158655

   c) 0.841345 - 0.001350 = 0.839995

   d) 0.998650 - 0.022750 = 0.9759

ANSWERS TO TESTING YOUR UNDERSTANDING OF THIS CHAPTER'S CONCEPTS - (PAGE 441)

1. The one with mean $\mu = 2$ and standard deviation $\sigma = 4$.

5. For all parts the probability is $0.3888 + 0.4788 = 0.8676$ because for a continuous distribution the probability that z equals a particular value is 0.

ANSWERS TO CHAPTER TEST - (PAGES 441 - 446)

1. $0.4306 + 0.4788 = 0.9094$ Choice (d)

5. $z = \dfrac{8.5 - 8}{1.2} = 0.42$
   $\begin{array}{r} 0.5000 \\ -0.1628 \\ \hline 0.3372 \end{array}$
   33.72%
   Choice (c)

9. $z = \dfrac{103 - 95}{7} = 1.14$

   $z = \dfrac{90 - 95}{7} = -0.71$
   $\begin{array}{r} 0.3729 \\ +0.2611 \\ \hline 0.6340 \end{array}$

13.

   $\mu = 100(0.35) = 35$

   $z = \dfrac{39.5 - 35}{4.7697} = 0.94$
   $\begin{array}{r} 0.5000 \\ -0.3264 \\ \hline 0.1736 \end{array}$

   $\sigma = \sqrt{100(0.35)(0.65)}$

   ≈ 4.7697

17. $\quad z = \dfrac{30,000 - 35,000}{2200} = -2.27 \quad \begin{array}{r} 0.5000 \\ -0.4884 \\ \hline 0.0116 \end{array}$

21.

$$\mu = 250(0.53) = 132.5 \qquad z = \frac{130.5 - 132.5}{7.8915} = -0.25 \quad \begin{array}{r} 0.5000 \\ -0.0987 \\ \hline 0.4013 \end{array}$$

$$\sigma = \sqrt{250(0.53)(0.47)}$$

$$\approx 7.8915$$

25.

$$\mu = 150(0.70) = 105 \qquad z = \frac{100.5 - 105}{5.6125} = -0.80 \quad \begin{array}{r} 0.5000 \\ -0.2881 \\ \hline 0.2119 \end{array}$$

$$\sigma = \sqrt{150(0.70)(0.30)}$$

$$\approx 5.6125$$

ANSWERS TO THINKING CRITICALLY - (PAGE 446)

1.

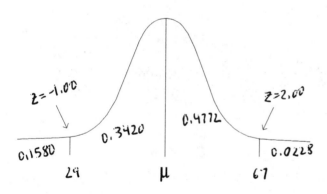

$29 = \mu - 1.00\sigma$
$67 = \mu + 2.00\sigma$

Solving simultaneously for μ and σ gives
$\mu = 41.6667 \quad \sigma = 12.6667$

5. When $n = 12$ and $p = 0.05$, we can use the binomial distribution formula.

p(at most 1 success) $= p$(0 successes) $+ p$(1 success)

$$= \frac{12!}{0!\,12!}\,(0.05)^0\,(0.95)^{12} + \frac{12!}{1!\,11!}\,(0.05)^1\,(0.95)^{11}$$

$$= 0.54036 + 0.34128 = 0.88164$$

Using normal approximation, we have

$$\mu = np\ 12(0.05) = 0.6 \qquad z = \frac{1.5 - 0.6}{0.7550} = 1.19$$

$$\begin{array}{r} 0.5000 \\ + 0.3830 \\ \hline 0.8830 \end{array}$$

$$\sigma = \sqrt{12(0.05)(0.95)}$$

$$\approx 0.7550$$

The answers are slightly different since np and nq are not greater than 5.

ANSWERS TO CASE STUDY - (PAGE 447)

1.

$$\mu = 537(0.487) = 261.519 \qquad z = \frac{230.5 - 261.519}{11.5827} = -2.68$$

$$\begin{array}{r} 0.5000 \\ - 0.4963 \\ \hline 0.0037 \end{array}$$

$$\sigma = \sqrt{537(0.487)(0.513)}$$

$$\approx 11.5827$$

CHAPTER 8

ANSWERS TO EXERCISES FOR SECTION 8.2

1. Those requests whose numbers are 691, 279, 151, 394, 604, 186, 711, 577, 388, 568, 186, 363, 676, 475, 607, 553, 185, 458, 609, and 665.

5. Those customers whse numbers are 14194, 53402, 24830, 53537, 81305, 70659, 18738, 56869, 84378, 62300, 05859, 72695, 17617, 81056, 92144, 44819, 29852, 13602, 04734, 26384, 28728, 15398, 61280, 14778, 81536, 61362, 63904, 22209, 36086, 08625, 82271, 35797, 20542, 58727, and 25417.

9. Those volunteers whose numbers are 0201, 1665, 0797, 1028, 0342, 0817, 0999, 1434, 2420, 0735, 2643, 2642, 1290, 3013, 0739, 2999, 3192, 2538, 3099, and 0785.

ANSWERS TO EXERCISES FOR SECTION 8.5

1.

Number of patients, x	Sample means $\bar{x}$	$\bar{x} - \mu_{\bar{x}}$	$(\bar{x} - \mu_{\bar{x}})^2$
6 and 10	8	8 - 7.3333 = 0.6667	0.4445
6 and 4	5	5 - 7.3333 = -2.3333	5.4443
6 and 7	6.5	6.5 - 7.3333 = -0.8333	0.6944
6 and 9	7.5	7.5 - 7.3333 = 0.1667	0.0278
6 and 8	7	7 - 7.3333 = -0.3333	0.1111
10 and 4	7	7 - 7.3333 = -0.3333	0.1111
10 and 7	8.5	8.5 - 7.3333 = 1.1667	1.3612
10 and 9	9.5	9.5 - 7.3333 = 2.1667	4.6946
10 and 8	9	9 - 7.3333 = 1.6667	2.7779
4 and 7	5.5	5.5 - 7.3333 = -1.8333	3.3610
4 and 9	6.5	6.5 - 7.3333 = -0.8333	0.6944
4 and 8	6	6 - 7.3333 = -1.3333	1.7777
7 and 9	8	8 - 7.3333 = 0.6667	0.4445
7 and 8	7.5	7.5 - 7.3333 = 0.1667	0.0278
9 and 8	8.5	8.5 - 7.3333 = 1.1667	1.3612
	110.0		23.3335

c) $\mu_{\bar{x}} = \dfrac{110}{15} = 7.3333$ **d)** $\sigma_{\bar{x}} = \sqrt{\dfrac{23.3335}{15}} = \sqrt{1.5556} \approx 1.2472$

5. a)

x	$x - \mu$	$(x - \mu)^2$
12	$12 - 18 = -6$	36
21	$21 - 18 = 3$	9
16	$16 - 18 = -2$	4
15	$15 - 18 = -3$	9
26	$26 - 18 = 8$	64
90	0	122

$$\mu = \frac{90}{5} = 18$$

$$\sigma = \sqrt{\frac{122}{5}} = \sqrt{24.4} \approx 4.9396$$

b) <u>Size 2</u>

Number of cases, x	Sample means, $\bar{x}$	$\bar{x} - \mu_{\bar{x}}$	$(\bar{x} - \mu_{\bar{x}})^2$
12 and 21	16.5	-1.5	2.25
12 and 16	14.0	-4.0	16.00
12 and 15	13.5	-4.5	20.25
12 and 26	19.0	1.0	1.00
21 and 16	18.5	0.5	0.25
21 and 15	18.0	0.0	0.00
21 and 26	23.5	5.5	30.25
16 and 15	15.5	-2.5	6.25
16 and 26	21.0	3.0	9.00
15 and 26	20.5	2.5	6.25
	180.0		91.50

ANSWERS TO EXERCISES FOR SECTION 8.5

Size 3

Number of cases, x	Sample means, $\bar{x}$	$\bar{x} - \mu_{\bar{x}}$	$(\bar{x} - \mu_{\bar{x}})^2$
12, 21, and 16	16.3333	-1.6667	2.7779
12, 21, and 15	16.0000	-2.0000	4.0000
12, 21, and 26	19.6667	1.6667	2.7779
12, 16, and 15	14.3333	-3.6667	13.4447
12, 16, and 26	18.0000	0.0000	0.0000
12, 15, and 26	17.6667	-0.3333	0.1111
21, 16, and 15	17.3333	-0.6667	0.4445
21, 16, and 26	21.0000	3.0000	9.0000
21, 15, and 26	20.6667	2.6667	7.1113
16, 15, and 26	19.0000	1.0000	1.0000
	180.0000		40.6674

c) **Size 2**

$$\mu_{\bar{x}} = \frac{180}{10} = 18$$

$$\sigma_{\bar{x}} = \sqrt{\frac{91.5}{10}} = \sqrt{9.15} \approx 3.0249$$

Size 3

$$\mu_{\bar{x}} = \frac{180}{10} = 18$$

$$\sigma_{\bar{x}} = \sqrt{\frac{40.6674}{10}} = \sqrt{4.06674} \approx 2.0166$$

ANSWERS TO EXERCISES FOR SECTION 8.5

d) We use $\sigma_{\bar{x}} = \dfrac{\sigma}{\sqrt{n}} \sqrt{\dfrac{N-n}{N-1}}$. In both cases $N = 5$

$$\sigma_{\bar{x}} = \frac{4.9396}{\sqrt{2}} \sqrt{\frac{5-2}{5-1}} \quad \text{and} \quad \sigma_{\bar{x}} = \frac{4.9396}{\sqrt{3}} \sqrt{\frac{5-3}{5-1}}$$

$$\approx 3.0249 \qquad \text{and} \qquad \approx 2.0166$$

ANSWER TO EXERCISES FOR SECTION 8.7

1.

$$z = \frac{200 - 210}{38/\sqrt{64}} = -2.11$$

$$\begin{array}{r} 0.4826 \\ +\ 0.4992 \\ \hline 0.9818 \end{array}$$

$$z = \frac{225 - 210}{38/\sqrt{64}} = 3.16$$

5. $z = \dfrac{6.1 - 6.8}{2.3/\sqrt{60}} = -2.36$

$$\begin{array}{r} 0.5000 \\ +0.4909 \\ \hline 0.9909 \end{array}$$

9. $z = \dfrac{675 - 680}{18/\sqrt{45}} = -1.86$

$$\begin{array}{r} 0.5000 \\ +0.4686 \\ \hline 0.9686 \end{array}$$

ANSWER TO EXERCISES FOR SECTION 8.8

1.   MTB > RANDOM 10 OBS, into C1;
         SUBC > NORMAL MU = 75 , SIGMA = 10.
         MTB > AVERAGE THE OBSERVATIONS IN C1

Repeat the above procedure 50 times to obtain 50 sample means. These are then entered in C51.

ANSWERS TO EXERCISES FOR SECTION 8.8

MTB > SET THE FOLLOWING DATA INTO C51
DATA > ...
DATA > END
MTB > DESCRIBE C51

The results will vary depending upon the numbers generated. For an alternate and time-saving way to generate all 50 samples at the same time, see the last comment at the end of the section.

5.   Same as Exercise 1 except that we now have

MTB > RANDOM 15 OBS, into C1;
SUBC > NORMAL MU = 18 , SIGMA = 4.

Repeat above procedure 80 times. (See last comment at end of section.)

ANSWERS TO TESTING YOUR UNDERSTANDING OF THIS CHAPTER'S CONCEPTS - (484 - 485)

1. The relationship is expressed in the formula $\sigma_{\bar{x}} = \dfrac{\sigma}{\sqrt{n}}$ when the sample is less than

5% of the population size.

ANSWERS TO CHAPTER TEST - (PAGES 485 - 487)

1. $z = \dfrac{495 - 488}{18.3 / \sqrt{49}} = 2.68$

$$\begin{array}{r} 0.5000 \\ -\ 0.4963 \\ \hline 0.0037 \end{array}$$

Choice (a)

5.

$$1.96 = \frac{\bar{x} - 8.3}{2.13/\sqrt{36}} = 8.9958$$

$$-1.96 = \frac{\bar{x} - 8.3}{2.13/\sqrt{36}} = 7.6042$$

Between 7.6042 and 8.9958   Choice (c)

9. $z = \dfrac{14 - 15}{2.25/\sqrt{39}} = -2.78$

$$\begin{array}{r} 0.5000 \\ -\,0.4973 \\ \hline 0.0027 \end{array}$$

13.

$$z = \frac{80 - 84}{9/\sqrt{40}} = -2.81$$

$$z = \frac{87 - 84}{9/\sqrt{40}} = 2.11$$

$$\begin{array}{r} 0.4975 \\ +\,0.4826 \\ \hline 0.9801 \end{array}$$

17.

x	$x - \mu$	$(x - \mu)^2$
710	-48	2304
800	42	1764
740	-18	324
760	2	4
780	22	484
3790		4880

$$\sigma_{\bar{x}} = \sqrt{\frac{4880}{5}} = \sqrt{976} \approx 31.2410$$

21. Those doctors whose numbers are 691, 279, 151, 394, 604, 186, 711, 577, 388, and 568.

25. $\mu_{\bar{x}} = \dfrac{410 + 510 + 480 + 500 + 490}{5} = \dfrac{2390}{5} = 478$

29. $\mu_{\bar{x}} = 193 \qquad \sigma_{\bar{x}} = \dfrac{28}{\sqrt{100}} = 2.8$

ANSWERS TO THINKING CRITICALLY - (PAGES 487 - 488)

1. The larger the sample size the better the estimate of the population mean since it is based on more data which encompasses more of the population.

5. Probably yes.

ANSWERS TO CASE STUDY - (PAGES 488 - 489)

1. a)

$$1.96 = \dfrac{\bar{x} - 4.43}{1.02/\sqrt{81}} = 4.6521$$

$$-1.96 = \dfrac{\bar{x} - 4.43}{1.02/\sqrt{81}} = 4.2079$$

Between 4.2079 and 4.6521 hours.

b)

$$1.96 = \dfrac{\bar{x} - 1.98}{0.66/\sqrt{81}} = 2.1237$$

$$-1.96 = \dfrac{\bar{x} - 1.98}{0.66/\sqrt{81}} = 1.8363$$

Between 1.8363 and 2.1237 hours.

c) The interval for the average number of hours spent by 95% of urban young people listening to music is considerably larger than the interval for the average number of hours spent by 95% of suburban teenagers.

CHAPTER 9

ANSWERS TO EXERCISES FOR SECTION 9.3

1. Lower boundary: $128,670 - 1.645\left(\dfrac{8473}{\sqrt{100}}\right) = 127,276.19$

Upper boundary: $128,670 + 1.645\left(\dfrac{8473}{\sqrt{100}}\right) = 130,063.81$

90% confidence interval: $127,276.19 to $130,063.81

5. Lower boundary: $12 - 1.96\left(\dfrac{2.35}{\sqrt{49}}\right) = 11.34$

Upper boundary: $12 + 1.96\left(\dfrac{2.35}{\sqrt{49}}\right) = 12.66$

95% confidence interval: $11.34 to $12.66

9. Lower boundary: $2.2 - 1.645\left(\dfrac{0.57}{\sqrt{55}}\right) = 2.0736$

Upper boundary: $2.2 + 1.645\left(\dfrac{0.57}{\sqrt{55}}\right) = 2.3264$

90% confidence interval: 2.0736 to 2.3264 smoke detectors

ANSWERS TO EXERCISES FOR SECTION 9.4

1. Lower boundary: $\quad 39.95 - 2.365\left(\dfrac{2.85}{\sqrt{8}}\right) = \$37.57$

Upper boundary: $\quad 39.95 + 2.365\left(\dfrac{2.85}{\sqrt{8}}\right) = \$42.33$

95% confidence interval: $\quad \$37.57$ and $\$42.33$

5. We must first find $\bar{x}$ and s. We have

x	$x - \bar{x}$	$(x - \bar{x})^2$
170	23	529
180	33	1089
149	2	4
132	-15	225
149	2	4
140	-7	49
167	20	400
110	-37	1369
126	-21	441
1323		4110

$$\bar{x} = \frac{1323}{9} = 147$$

$$s = \sqrt{\frac{4110}{9-1}} = \sqrt{513.75} \approx 22.6661$$

Lower boundary : $147 - 1.860\left(\dfrac{22.6661}{\sqrt{9}}\right) = 132.9470$

Upper boundary: $147 + 1.860\left(\dfrac{22.6661}{\sqrt{9}}\right) = 161.0530$

90% confidence interval: Between 132.9470 and 161.0530 calories.

9. We must first find $\bar{x}$ and s. We have

x	$x - \bar{x}$	$(x - \bar{x})^2$
10	2.5	6.25
8	0.5	0.25
3	-4.5	20.25
2	-5.5	30.25
12	4.5	20.25
3	-4.5	20.25
11	3.5	12.25
14	6.5	42.25
7	-0.5	0.25
5	-2.5	6.25
75		158.50

$$\bar{x} = \frac{75}{10} = 7.5$$

$$s = \sqrt{\frac{158.50}{10 - 1}} = \sqrt{17.611} \approx 4.1966$$

ANSWERS TO EXERCISES FOR SECTION 9.4

Lower boundary : $7.5 - 1.833\left(\dfrac{4.1966}{\sqrt{10}}\right) = 5.0675$

Upper boundary: $7.5 + 1.833\left(\dfrac{4.1966}{\sqrt{10}}\right) = 9.9325$

90% confidence interval: Between 5.0675 and 9.9325 hours.

ANSWERS TO EXERCISES FOR SECTION 9.6

1.   Lower boundary : $\dfrac{9.75}{1 + \dfrac{1.96}{\sqrt{2(50)}}} = \$8.15$

   Upper boundary: $\dfrac{9.75}{1 - \dfrac{1.96}{\sqrt{2(50)}}} = \$12.13$

   95% confidence interval for σ: Between \$8.15 and \$12.13

5. $n = \left(\dfrac{1.96\,(425)}{100}\right)^2 = 69.3889$ Sample size = 70

9. $n = \left(\dfrac{1.96\,(24.75)}{20}\right)^2 = 5.8831$ Sample size = 6

ANSWERS TO EXERCISES FOR SECTION 9.7

1. a)

Lower Boundary	Upper Boundary
$= 0.45 - 1.96\sqrt{\dfrac{0.45(1-0.45)}{800}}$	$= 0.45 + 1.96\sqrt{\dfrac{0.45(1-0.45)}{800}}$
$= 0.45 - 1.96\sqrt{0.0003}$	$= 0.45 + 1.96\sqrt{0.0003}$
$= 0.45 - 1.96(0.0176)$	$= 0.45 + 1.96(0.0176)$
$= 0.45 - 0.0345$	$= 0.45 + 0.0345$
$= 0.4155$	$= 0.4845$

95% confidence interval: Between 0.4155 and 0.4845

b)

Lower Boundary	Upper Boundary
$= 0.65 - 1.96\sqrt{\dfrac{0.65(1-0.65)}{400}}$	$= 0.65 + 1.96\sqrt{\dfrac{0.65(1-0.65)}{400}}$
$= 0.65 - 1.96\sqrt{0.00057}$	$= 0.65 + 1.96\sqrt{0.00057}$
$= 0.65 - 1.96(0.0239)$	$= 0.65 + 1.96(0.0239)$
$= 0.65 - 0.0468$	$= 0.65 + 0.0468$
$= 0.6032$	$= 0.6968$

95% confidence interval: Between 0.6032 and 0.6968

5. $\hat{p} = \dfrac{1}{10} = 0.10$

Lower Boundary	Upper Boundary
$= 0.10 - 1.645\sqrt{\dfrac{0.10(1-0.10)}{75}}$	$= 0.10 + 1.645\sqrt{\dfrac{0.10(1-0.10)}{75}}$
$= 0.10 - 0.0570$	$= 0.10 + 0.0570$
$= 0.0430$	$= 0.1570$

90% confidence interval: Between 0.0430 and 0.1570

9.

$$\sigma_{\hat{p}} = \sqrt{\dfrac{0.33(1-0.33)}{275}} = 0.0284$$

$$\begin{array}{r} 0.5000 \\ -\,0.3554 \\ \hline 0.1446 \end{array}$$

$$z = \dfrac{0.30 - 0.33}{0.0284} = -1.06$$

ANSWERS TO EXERCISES FOR SECTION 9.8

1.   MTB > SET THE FOLLOWING DATA IN C1
   DATA >   23 17 16 8 31 14 12 17 15 17 15 13
   DATA >   END
   MTB > TINTERVAL WITH 90 PERCENT CONFIDENCE FOR DATA IN C1

Confidence Intervals

Variable	N	Mean	StDev	SE Mean	90.0 % C.I.	
C1	12	16.50	5.79	1.67	(13.50,	19.50)

ANSWERS TO TESTING YOUR UNDERSTANDING OF THIS CHAPTER'S CONCEPTS - (PAGE 525)

1. Choice (d)

ANSWERS TO CHAPTER TEST - (PAGES 525 - 528)

1.   Lower boundary : $49.95 - 1.96\left(\dfrac{1.75}{\sqrt{50}}\right) = \$49.46$

Upper boundary: $49.95 + 1.96\left(\dfrac{1.75}{\sqrt{50}}\right) = \$50.44$

Between \$49.46 and \$50.44   Choice (a)

5.   Lower boundary : $\dfrac{69}{1 + \dfrac{2.58}{\sqrt{2(45)}}} = 54.2472$

Upper boundary: $\dfrac{69}{1 - \dfrac{2.58}{\sqrt{2(45)}}} = 94.7745$

Between   \$54.25 and \$94.77   Choice (b)

9. $\hat{p} = \dfrac{140}{250} = 0.56$

Lower Boundary	Upper Boundary
$= 0.56 - 1.96\sqrt{\dfrac{0.56\,(1 - 0.56)}{250}}$	$= 0.56 + 1.96\sqrt{\dfrac{0.56\,(1 - 0.56)}{250}}$
$= 0.56 - 0.0615$	$= 0.56 + 0.0615$
$= 0.4985$	$= 0.6215$

95% confidence interval: Between 0.4985 and 0.6215

13.   Lower boundary : $120 - 1.860 \left(\dfrac{11}{\sqrt{9}} \right) = 113.18$

Upper boundary: $120 + 1.860 \left(\dfrac{11}{\sqrt{9}} \right) = 126.82$

90% confidence interval: Between \$113.18 and \$126.82

17.

$\sigma_{\hat{p}} = \sqrt{\dfrac{0.15\,(1 - 0.15)}{100}} = 0.0357$

$$\begin{array}{r} 0.5000 \\ - 0.2995 \\ \hline 0.2005 \end{array}$$

$z = \dfrac{0.12 - 0.15}{0.0357} = -0.84$

21.   Lower boundary : $21.95 - 1.96 \left(\dfrac{2.25}{\sqrt{45}} \right) = 21.2926$

Upper boundary: $21.95 + 1.96 \left(\dfrac{2.25}{\sqrt{45}} \right) = 22.6074$

95% confidence interval: Between \$21.29 and \$22.61

ANSWERS TO THINKING CRITICALLY - (PAGE 528)

1. True. To double accuracy, we cut e in half and the formula indicates that we need to quadruple the sample size.

5. Large sample size.

1. $\sigma_{\hat{p}} = \sqrt{\dfrac{0.142\,(1 \,-\, 0.142)}{600}} = 0.0142$

Lower Boundary	Upper Boundary
= 0.142 - 1.96(0.0142)	= 0.142 + 1.96(0.0142)
= 0.142 - 0.0278	= 0.142 + 0.0278
= 0.1142	= 0.1698

95% confidence interval: Between 0.1142 and 0.1698

CHAPTER 10

ANSWERS TO EXERCISES FOR SECTION 10.4

1. $z = \dfrac{34000 - 35000}{2400/\sqrt{50}} = -2.95$

    Reject manufacturer's claim

5. $z = \dfrac{84.93 - 82.86}{5.88/\sqrt{60}} = 2.73$

    Yes. We reject the claim that appears in <u>Statistical Abstract of the United States.</u>

    Average cost has risen.

9. $z = \dfrac{7.3 - 6.9}{1.3/\sqrt{37}} = 1.87$

    Yes. We cannot reject Motor Vehicle Bureau claim.

ANSWERS TO EXERCISES FOR SECTION 10.5

1. $t = \dfrac{9.3 - 8.7}{0.04/\sqrt{10}} = 47.43$

    Yes. The charge for electricity by these utilities is significantly above average.

5. $t = \dfrac{28.15 - 26.20}{1.09/\sqrt{9}} = 5.37$

    Yes. Reject claim that the average fare is \$26.20.

ANSWERS TO EXERCISES FOR SECTION 10.5

9. $t = \dfrac{52 - 57}{12 / \sqrt{12}} = -1.44$

Do not reject null hypothesis.

ANSWERS TO EXERCISES FOR SECTION 10.6

1. $z = \dfrac{1.38 - 1.31}{\sqrt{\dfrac{(0.04)^2}{41} + \dfrac{(0.09)^2}{53}}} = \dfrac{0.07}{0.0139} = 5.036$

Reject null hypothesis. There is a significant difference.

5. $z = \dfrac{12.55 - 11.12}{\sqrt{\dfrac{(2.12)^2}{45} + \dfrac{(1.89)^2}{38}}} = \dfrac{1.43}{0.4403} = 3.25$

Reject null hypothesis. There is a significant difference.

9. $z = \dfrac{7 - 8}{\sqrt{\dfrac{(1.86)^2}{60} + \dfrac{(2.16)^2}{75}}} = \dfrac{-1}{0.3462} = -2.89$

Reject null hypothesis. There is a significant difference.

ANSWERS TO EXERCISES FOR SECTION 10.7

1.

$s_p = \sqrt{\dfrac{(16 - 1)(3.26)^2 + (12 - 1)(8.88)^2}{16 + 12 - 2}} \approx 6.2843$

$t = \dfrac{18.8 - 28.6}{(6.2843)\sqrt{\dfrac{1}{16} + \dfrac{1}{12}}} = -4.08$

Reject null hypothesis. There is evidence to indicate a significant difference.

ANSWERS TO EXERCISES FOR SECTION 10.7

5.

$$s_p = \sqrt{\frac{(6 - 1)(1.86)^2 + (5 - 1)(2.13)^2}{6 + 5 - 2}} \approx 1.9845$$

$$t = \frac{7 - 6}{(1.9845)\sqrt{\frac{1}{6} + \frac{1}{5}}} = 0.8322$$

Do not reject null hypothesis. There is not enough information to indicate a significant difference.

9.

$$s_p = \sqrt{\frac{(10 - 1)(812)^2 + (16 - 1)(675)^2}{10 + 16 - 2}} \approx 729.3968$$

$$t = \frac{6375 - 5628}{(729.3968)\sqrt{\frac{1}{10} + \frac{1}{16}}} = 2.54$$

Reject null hypothesis. There is evidence to indicate a significant difference.

ANSWERS TO EXERCISES FOR SECTION 10.8

1.

$$\hat{p} = \frac{10}{84} = 0.119$$

$$z = \frac{0.119 - 0.10}{\sqrt{\frac{0.10(1 - 0.10)}{84}}} = 0.58$$

No. Do not reject null hypothesis and the study.

ANSWERS TO EXERCISES FOR SECTION 10.8

5.

$$\hat{p} = \frac{24}{80} = 0.30$$

$$z = \frac{0.30 - 0.25}{\sqrt{\dfrac{0.25(1 - 0.25)}{80}}} = 1.03$$

No. Do not reject null hypothesis and the claim.

9.

$$\hat{p} = \frac{35}{50} = 0.70$$

$$z = \frac{0.70 - 0.66}{\sqrt{\dfrac{0.66(1 - 0.66)}{50}}} = 0.5971$$

$$z = \frac{0.70 - 0.71}{\sqrt{\dfrac{0.71(1 - 0.71)}{50}}} = -0.1558$$

No. Results do not differ from IIHS or <u>Prevention</u> claim. Do not reject null hypothesis.

ANSWERS TO EXERCISES FOR SECTION 10.9

1.
```
MTB  >  SET THE FOLLOWING IN C1
DATA >  36  34  39  43  29  32  35  36
DATA >  END
MTB  >  TTEST , MU = 44,  C1
```

TEST OF MU = 44.000 VS MU NOT = 44.000

	N	MEAN	STDEV	SE MEAN	T	P VALUE
C1	8	35.50	4.24	1.50	-5.67	0.0008

ANSWERS TO EXERCISES FOR SECTION 10.9

Since the P-value of 0.0008 is less than 0.05, we must reject null hypothesis.

There is a significant difference between the average amount of pollutant in the air

found by health officials and the union claim.

5.   MTB > SET THE FOLLOWING IN C1
    DATA > 6 8 4 2 3 7 8 1 3 2 7 9
    DATA > END
    MTB > TTEST , MU = 8.62 , C1;
    SUBC > ALTERNATIVE = -1.

TEST OF MU = 8.620 VS MU < 8.620

	N	MEAN	STDEV	SE MEAN	T	P VALUE
C1	12	5.000	2.796	0.807	-4.48	0.0005

Since the P-value of 0.0005 is less than 0.05, we reject the null hypothesis and the

welfare department official's claim.

9.   MTB > SET THE FOLLOWING IN C1
    DATA > 24.6 25.3 28.6 26.1 23.7 26.4 25.9 25.1 24.3
    DATA > END
    MTB > SET THE FOLLOWING IN C2
    DATA > 26.4 27.3 25.4 26.1 24.2 23.8 24.0 22.1
    DATA > END
    MTB > TWOSAMPLE T C1 C2;
    SUBC > ALTERNATIVE = +1.

TWOSAMPLE T FOR C1 VS C2

	N	MEAN	STDEV	SE MEAN
C1	9	25.56	1.44	0.48
C2	8	24.91	1.69	0.60

95% C.I. FOR MU C1 - MU C2 : (-1.01, 2.30)
T-TEST MU C1 = MU C2 (VS >) : T = 0.84 P = 0.21 DF = 13

ANSWERS TO EXERCISES FOR SECTION 10.9

Since the P-value of 0.21 is greater than 0.05, we do not reject the null hypothesis and conclude that there is not sufficient evidence to say that the new chemical additive significantly increases gasoline mileage.

ANSWERS TO TESTING YOUR UNDERSTANDING OF THIS CHAPTER'S CONCEPTS - (PAGES 579 - 581)

1.

$$s_p = \sqrt{\frac{(12 - 1)(3.1)^2 + (14 - 1)(2.98)^2}{12 + 14 - 2}} \approx 3.0356$$

The endpoints of the confidence interval for $\mu_1 - \mu_2$ are

$$\left(\overline{x_1} - \overline{x_2}\right) \pm t_{\frac{\alpha}{2}} \cdot s_p \sqrt{\frac{1}{n_1} + \frac{1}{n_2}} \qquad df = n_1 + n_2 - 2 = 12 + 14 - 2 = 24$$

Lower boundary : $(25 - 11) - (1.711)(3.0356)\sqrt{\frac{1}{12} + \frac{1}{14}} = 11.96$

Upper boundary: $(25 - 11) + (1.711)(3.0356)\sqrt{\frac{1}{12} + \frac{1}{14}} = 16.04$

90% confidence interval: Between 11.96 cents and 16.04 cents.

ANSWERS TO CHAPTER TEST - (PAGES 580 - 584)

1. $z = \dfrac{475 - 495}{47/\sqrt{86}} = -3.95$ Reject null hypothesis. Company is paying lower wages.

Choice (a)

5. $z = \dfrac{25 - 36}{\sqrt{\dfrac{(2.7)^2}{80} + \dfrac{(3.9)^2}{111}}} = -23.029$ Reject null hypothesis. Choice (a)

9.

$\hat{p} = \dfrac{48}{58} = 0.8276$

$z = \dfrac{0.8276 - 0.85}{\sqrt{\dfrac{0.85(1 - 0.85)}{58}}} = -0.48$

Do not reject null hypothesis. Do not reject chemist's claim.

13. $t = \dfrac{36000 - 34000}{2200/\sqrt{8}} = 2.57$

Do not reject null hypothesis. The special mechanism does not significantly affect the

average life of a transmission.

17.

$s_p = \sqrt{\dfrac{(7 - 1)(4.69)^2 + (9 - 1)(6.13)^2}{7 + 9 - 2}} \approx 5.5587$

$t = \dfrac{58 - 63}{5.5587\sqrt{\dfrac{1}{7} + \dfrac{1}{9}}} = -1.78$

Do not reject null hypothesis. The difference in the average coliform of coliform bacteria

in both states is not statistically significant.

21.

$$z = \frac{7.1 - 6.8}{\sqrt{\frac{(1.48)^2}{55} + \frac{(1.96)^2}{65}}} = 0.95$$

Do not reject null hypothesis. The difference between the average lasting time of the paints is not statistically significant.

25.

$$\hat{p} = \frac{200}{300} = 0.6667$$

$$z = \frac{0.6667 - 0.70}{\sqrt{\frac{0.70(1 - 0.70)}{300}}} = -1.26$$

Do not reject null hypothesis. There is not sufficient evidence to reject the politician's claim.

ANSWERS TO THINKING CRITICALLY - (PAGE 584)

1. No. When the null hypothesis is rejected, either a type I error is made or a correct decision is made. When the null hypothesis is not rejected, either a type II error is made or a correct decision is made. Since it is not possible to simultaneously reject and accept the null hypothesis, it is not possible to make both errors at the same time.

1.

$$\hat{p} = \frac{33}{98} = 0.3367$$

$$z = \frac{0.3367 - 0.23}{\sqrt{\dfrac{0.23(1 - 0.23)}{98}}} = 2.51$$

Reject null hypothesis. The number of births by Caesarean section is significantly more than the claimed 23%.

CHAPTER 11

ANSWERS TO EXERCISES FOR SECTION 11.3

1. **a)** Positive correlation

   b) Zero correlation

   c) Probably positive correlation. Others may disagree.

   d) Positive correlation

   e) Positive correlation

   f) Negative correlation

   g) Positive correlation

   h) Positive correlation

5.

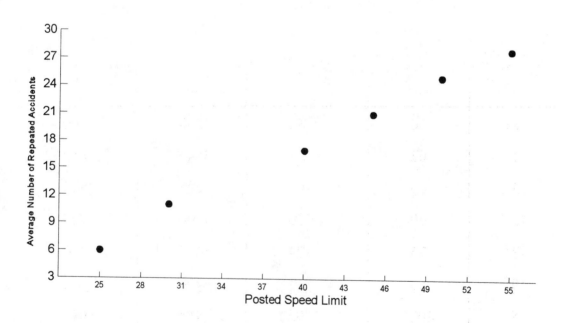

x	y	x^2	y^2	xy
55	28	3025	784	1540
50	25	2500	625	1250
45	21	2025	441	945
40	17	1600	289	680
30	11	900	121	330
25	6	625	36	150
245	108	10675	2296	4895

$$r = \frac{6(4895) - (245)(108)}{\sqrt{6(10675) - 245^2}\ \sqrt{6(2296) - 108^2}}$$

$$= \frac{2910}{\sqrt{4025}\ \sqrt{2112}} = 0.9981$$

9.

x	y	x^2	y^2	xy
8	50	64	2500	400
11	62	121	3844	682
13	65	169	4225	845
16	70	256	4900	1120
18	75	324	5625	1350
19	80	361	6400	1520
85	402	1295	27,494	5917

ANSWERS TO EXERCISES FOR SECTION 11.3

$$r = \frac{6(5917) - (85)(402)}{\sqrt{6(1295) - 85^2} \ \sqrt{6(27494) - 402^2}}$$

$$= \frac{1332}{\sqrt{545} \ \sqrt{3360}} = 0.9843$$

ANSWERS TO EXERCISES FOR SECTION 11.4

1. Significant

5. Significant

9. Significant

ANSWERS TO EXERCISES FOR SECTION 11.6

1.

x	y	x^2	y^2	xy
64	66	4096	4356	4224
65	67	4225	4489	4355
68	69	4624	4761	4692
73	74	5329	5476	5402
72	73	5184	5329	5256
66	68	4356	4624	4488
71	72	5041	5184	5112
75	75	5625	5625	5625
70	71	4900	5041	4970
69	69	4761	4761	4761
693	704	48141	49646	48885

$$b_1 = \frac{10(48885) - (693)(704)}{10(48141) - 693^2} = \frac{978}{1161} = 0.84238$$

$$b_0 = \frac{1}{10}\left[704 - (0.84238)(693)\right] = 12.023$$

a) Least-squares prediction equation: $\hat{y} = 12.023 + 0.84238x$

b) $\hat{y} = 12.023 + 0.84238(74) = 74.359$ inches

5.

x	y	x^2	y^2	xy
35	1.6	1225	2.56	56.0
37	2.2	1369	4.84	81.4
39	3.8	1521	14.44	148.2
42	4.3	1764	18.49	180.6
44	5.6	1936	31.36	246.4
46	6.1	2116	37.21	280.6
50	7.3	2500	53.29	365.0
293	30.9	12431	162.19	1358.2

$$b_1 = \frac{7(1358.2) - (293)(30.9)}{7(12431) - 293^2} = \frac{453.7}{1168} = 0.38844$$

$$b_0 = \frac{1}{7}\left(30.9 - (0.38844)(293)\right) = -11.845$$

a) Least - squares prediction equation: $\hat{y} = -11.845 + 0.3884x$

b) $\hat{y} = -11.845 + 0.3884(48) = 6.800$ accidents.

9.

x	y	x^2	y^2	xy
3.89	0.39	15.1321	0.1521	1.5171
7.67	0.57	58.8289	0.3249	4.3719
8.19	0.59	67.0761	0.3481	4.8321
8.57	0.62	73.4449	0.3844	5.3134
9.00	0.63	81.0000	0.3969	5.6700
12.59	0.86	158.5081	0.7396	10.8274
21.59	1.19	466.1281	1.4161	25.6921
31.77	1.33	1009.3329	1.7689	42.2541
28.52	1.22	813.3904	1.4884	34.7944
26.19	1.16	685.9161	1.3456	30.3804
25.88	1.13	669.7744	1.2769	29.2444
42.09	1.12	1771.5681	1.2544	47.1408
12.51	0.86	156.5001	0.7396	10.7586
15.41	0.90	237.4681	0.8100	13.8690
253.87	12.57	6264.0683	12.4459	266.6657

$$b_1 = \frac{14(266.6657) - (253.87)(12.57)}{14(6264.0683) - (253.87)^2} = \frac{542.1739}{23246.9793} = 0.023322$$

$$b_0 = \frac{1}{14}\left[12.57 - (0.023322)(253.87)\right] = 0.47494$$

a) Least - squares prediction equation: $\hat{y} = 0.47494 + 0.023322x$

b) $\hat{y} = 0.47494 + (0.023322)(16.50) = 0.8598$

ANSWERS TO EXERCISES FOR SECTION 11.7

1.

$$s_e = \sqrt{\frac{49646 - (12.023)(704) - (0.84238)(48885)}{10 - 2}} \approx 0.5019$$

5.

$$s_e = \sqrt{\frac{162.19 - (-11.845)(30.9) - (0.38844)(1358.2)}{7 - 2}} \approx 0.3499$$

9.

$$s_e = \sqrt{\frac{12.4459 - (0.47494)(12.57) - (0.023322)(266.6657)}{14 - 2}} \approx 0.1462$$

ANSWERS TO EXERCISES FOR SECTION 11.8

1.

$$t = \frac{0.8424}{0.50\sqrt{48141 - \frac{693^2}{10}}} \qquad\qquad t_{0.025} = 2.306$$

$$= 18.1537$$

Reject H_0: $\beta_1 = 0$

When $x_p = 74$, $\hat{y}_p = 74.359$

Also $s_e = 0.5019$ and $\bar{x} = \frac{693}{10} = 69.3$

Lower boundary $= 74.359 - (2.306)(0.5019) \sqrt{1 + \dfrac{1}{10} + \dfrac{10(74 - 69.3)^2}{10(48141) - 693^2}}$

$= 74.359 - (2.306)(0.5019) \sqrt{1 + 0.1 + 0.190267011}$

$= 74.359 - (2.306)(0.5019)(1.135899208)$

$= 74.359 - 1.314668616 = 73.044$

Upper boundary $= 74.359 + (2.306)(0.5019) \sqrt{1 + 0.1 + 0.190267011}$

$= 74.359 + 1.314668616 = 75.674$

95% prediction interval :   73.044 to 75.674

5.

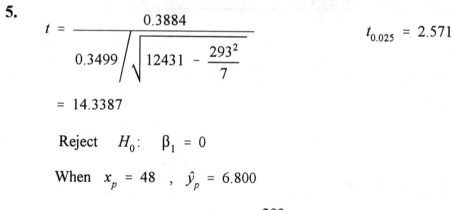

$$t = \frac{0.3884}{0.3499 \Big/ \sqrt{12431 - \dfrac{293^2}{7}}} \qquad\qquad t_{0.025} = 2.571$$

$= 14.3387$

Reject $H_0:$ $\beta_1 = 0$

When $x_p = 48$, $\hat{y}_p = 6.800$

Also $s_e = 0.3499$ and $\bar{x} = \dfrac{293}{7} = 41.857142$

Lower boundary $= 6.800 - (2.571)(0.3499) \sqrt{1 + \dfrac{1}{7} + \dfrac{7(48 - 41.847142)^2}{7(12431) - 293^2}}$

$= 6.800 - (2.571)(0.3499) \sqrt{1 + 0.1428571 + 0.2261497}$

$= 6.800 - (2.571)(0.3499) \sqrt{1.3690068}$

$= 6.800 - (2.571)(0.3499)(1.1700456)$

$= 6.800 - 1.0525647 = 5.748$

Upper boundary $= 6.800 + (2.571)(0.3499) \sqrt{1 + 0.1428571 + 0.2261497}$

$= 6.800 + 1.0525647 = 7.853$

95% prediction interval :   5.748 to 7.853

9.

$t = \dfrac{0.0233}{0.1462 \bigg/ \sqrt{6264.1363 - \dfrac{(253.87)^2}{14}}}$ $t_{0.025} = 2.179$

$= 6.4944$

Reject H_0 : $\beta_1 = 0$

When $x_p = 16.50$, $\hat{y}_p = 0.8598$

Also $s_e = 0.1462$ and $\bar{x} = \dfrac{253.87}{14} = 18.133571$

113

Lower boundary $= 0.8598 - (2.179)(0.1462)\sqrt{1 + \dfrac{1}{14} + \dfrac{14(1650 - 18.133571)^2}{14(6264.1363) - (253.87)^2}}$

$= 0.8598 - (2.179)(0.1462)\sqrt{1 + 0.0714285 + 0.001607}$

$= 0.8598 - (2.179)(0.1462)\sqrt{1.0730355}$

$= 0.8598 - (2.179)(0.1462)(1.0358742)$

$= 0.5296$

Upper boundary $= 0.8598 + (2.179)(0.1462)\sqrt{1 + 0.0714285 + 0.001607}$

$= 1.1899$

95% prediction interval: 0.5296 to 1.1899

ANSWERS TO EXERCISES FOR SECTION 11.10

1.

x_1	x_2	y	$x_1 y$	$x_2 y$	x_1^2	$x_1 x_2$	x_2^2
52	170	130	6760	22100	2704	8840	28900
53	175	135	7155	23625	2809	9275	30625
56	180	140	7840	25200	3136	10080	32400
58	186	145	8410	26970	3364	10788	34596
60	195	150	9000	29250	3600	11700	38025
63	200	155	9765	31000	3969	12600	40000
65	208	160	10400	33280	4225	13520	43264
70	215	165	11550	35475	4900	15050	46225
477	1529	1180	70880	226900	28707	91853	294035

ANSWERS TO EXERCISES FOR SECTION 11.10

$$1180 = 8b_0 + 477b_1 + 1529b_2$$

$$70880 = 477b_0 + 28707b_1 + 91853b_2$$

$$226900 = 1529b_0 + 91853b_1 + 294035b_2$$

Solving simultaneously gives $b_0 = 3.648$

$$b_1 = 0.1139 \quad b_2 = 0.7171$$

a) The least-squares prediction equation is $\hat{y} = 3.648 + 0.1139x_1 + 0.7171x_2$

b) $\hat{y} = 3.648 + 0.1139(67) + 0.7171(205) = 158.290$

ANSWERS TO EXERCISES FOR SECTION 11.11

1.
```
MTB  >  READ AGE IN C1, PERCENT REDUCTION IN C2
DATA >  52  60
DATA >  53  54
DATA >  55  46
DATA >  56  40
DATA >  57  35
DATA >  58  23
DATA >  59  18
DATA >  60  12
DATA >  61   4
DATA > END
MTB  NAME  C1  'x'  C2  'y'
MTB  PLOT  'y' vs 'x'
```

Character Plot

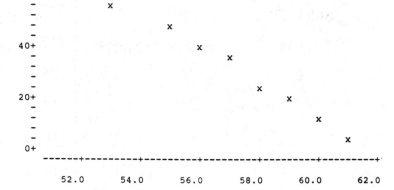

ANSWERS TO EXERCISES FOR SECTION 11.11

MTB > CORRELATION 'y' 'x'

Correlations (Pearson)

Correlation of y and x = -0.992

MTB > REGRESS 'y' 1 'x';
SUBC > PREDICT 54.

Regression Analysis

The regression equation is
y = 387 - 6.25 x

Predictor	Coef.	Stdev	t-ratio	p
Constant	387.22	16.81	23.04	0.000
x	-6.2485	0.2956	-21.14	0.000

s = 2.570   R-sq = 98.5%   R-sq(adj) = 98.2%

Analysis of Variance

SOURCE	DF	SS	MS	F	p
Regression	1	2950.0	2950.0	446.76	0.000
Error	7	46.2	6.6		
Total	8	2996.2			

Fit	Stdev. Fit	95.0% C.I.	95.0% P.I.
49.801	1.187	(46.995, 52.608)	(43.107, 56.496)

ANSWERS TO TESTING YOUR UNDERSTANDING OF THIS CHAPTER'S CONCEPTS - (PAGE 647)

1. Most likely choice (d). However, it really can be large, small, or anything in between as it will be approximately equal to the standard deviation of the y's.

5. a)

x	y	x^2	xy
8	70	64	560
6	60	36	360
3	50	9	150
9	82	81	738
15	99	225	1485
12	90	144	1080
53	451	559	4373

$$b_1 = \frac{6(4373) - (53)(451)}{6(559) - 53^2} = \frac{2335}{545} = 4.2844$$

$$b_0 = \frac{1}{6}\left[451 - (4.2844)53\right] = 37.3211$$

Least - squares prediction equation: $\hat{y} = 37.3211 + 4.2844x$

b) $75 = 37.3211 + 4.2844x$ Solving for x gives $x = 8.7944$ units of the drug.

ANSWERS TO CHAPTER TEST - (PAGES 647 - 651)

1.

x	y	x^2	y^2	xy	$y - \bar{y}$	$(y - \bar{y})^2$	$\hat{y}$	$y - \hat{y}_p$
0	16000	0	256,000,000	0	8083.3334	65,340,288	14,184.52	1815.48
1	11000	1	121,000,000	11,000	3083.3334	9,506,948.5	12,095.52	-1095.234
2	9000	4	81,000,000	18,000	1083.3334	1,173,611.2	10,005.948	-1005.948
4	5000	16	25,000,000	20,000	-2916.6667	8,506,944	5827.376	-827.376
5	4000	25	16,000,000	20,000	-3916.6667	15,340,278	3738.09	261.91
6	2500	36	6,250,000	15,000	-5416.6667	29,340,278	1648.804	851.196
18	47500	82	505,250,000	84,000		129,208,347.7		

$$r = \frac{6(84000) - 18(47500)}{\sqrt{6(82) - 18^2} \ \sqrt{6(505,250,000) - 47500^2}} = -0.973 \quad \text{Choice (a)}$$

5.

$$s_e = \sqrt{\frac{(505,250,000) - (14184.5238)(47500) - (-2089.2857)(84000)}{6 - 2}} \approx 1321.469$$

Choice (d)

9. Choice (c)

13. $\hat{y} = 69.2143 - 16.2286(2.25) = 32.69995$

17. Yes

21. a) Character Plot

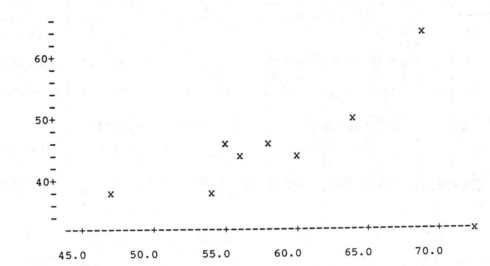

b)

x	y	x2	y2	xy
58	45	3364	2025	2610
54	37	2916	1369	1998
47	38	2209	1444	1786
60	44	3600	1936	2640
56	43	3136	1849	2408
55	45	3025	2025	2475
64	50	4096	2500	3200
69	63	4761	3969	4347
463	365	27107	17117	21464

$$r = \frac{8(21464) - (463)(365)}{\sqrt{8(27107) - 463^2}\ \sqrt{8(17117) - 365^2}} = 0.8943$$

119

c)

$$b_1 = \frac{8(21464) - (463)(365)}{8(27107) - 463^2} = \frac{2717}{2487} = 1.0925$$

$$b_0 = \frac{1}{8}\left[365 - (1.0925)463\right] = -17.6023$$

Least - squares prediction equation: $\hat{y} = -17.6023 + 1.0925x$

d) $\hat{y} = -17.6023 + 1.0925(62) = 50.1327$

e)

$$s_e = \sqrt{\frac{17117 - (-17.6023)(365) - (1.0925)(21464)}{8 - 2}} \approx 3.934$$

$t_{0.025} = 2.447$ and $\bar{x} = \frac{463}{8} = 57.875$ When $x_p = 62$ $\hat{y} = 50.1327$

Lower boundary $= 50.1327 - (2.447)(3.934)\sqrt{1 + \frac{1}{8} + \frac{8(62 - 57.875)^2}{8(27107) - 463^2}}$

$\qquad = 50.1327 - (2.447)(3.934)\sqrt{1 + 0.125 + 0.054734}$

$\qquad = 50.1327 - (2.447)(3.934)\sqrt{1.1797346}$

$\qquad = 50.1327 - 10.455877 = 39.6768$

Upper boundary $= 50.1327 + (2.447)(3.934)\sqrt{1.1797346}$

$\qquad = 50.1327 + 10.455877 = 60.5886$

95% prediction interval :   39.6768 to 60.5886

ANSWERS TO THINKING CRITICALLY - (PAGES 651 - 652)

1. Start with the formula for b_1 in Formula 11.2. Divide numerator and denominator by n - 1. Using the fact that $s_x^2 = \dfrac{n\Sigma x^2 - (\Sigma x)^2}{n(n-1)}$ we get the equation for b_1 given in

   Formula 11.3. Also the formulas for b_0 are equivalent since

   $\dfrac{1}{n}\Sigma x = \bar{x}$ and $\dfrac{1}{n}\Sigma y = \bar{y}$

ANSWERS TO CASE STUDIES - (PAGES 652 - 653)

1. a) Character Plot

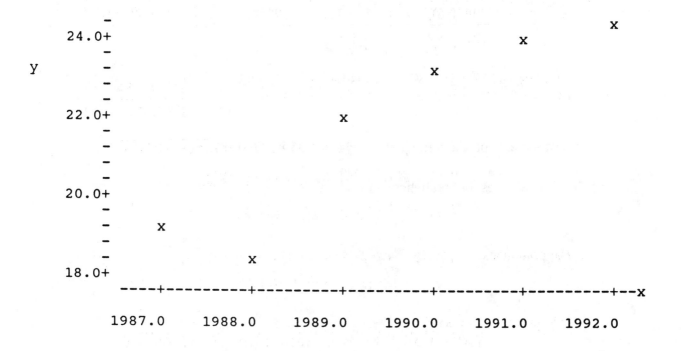

b)

x	y	$x2$	xy
1989	22.0	3956121	43758.0
1990	23.1	3960100	45969.0
1991	23.9	3964081	47584.9
1992	24.4	3968064	48604.8
1993	25.9	3972049	51618.7
1994	26.3	3976036	52442.2
1995	27.4	3980025	54663.0
13944	173.0	27,776,476	344,640.6

$$b_1 = \frac{7(344640.6) - (13944)(173)}{7(27,776,476) - 13944^2} = \frac{172.2}{196} = 0.8786$$

$$b_0 = \frac{1}{7}\left[170 - (0.8786)(13944)\right] = -1725.8855$$

Least-squares prediction equation: $\hat{y} = -1725.8855 + 0.8786x$

c) Technically yes, but realistically no.

CHAPTER 12

ANSWERS TO EXERCISES FOR SECTION 12.2

<u>Note</u>: Although expected frequencies will usually be given to 4 decimal places and the chi-square subtotals to three decimal places, the calculations will usually be done using full calculator accuracy.

1.

	Age 20-29	Age 30-39	Age 40-49	Age over 50
Yes	167 (150.97)	131 (126.79)	49 (61.07)	25 (33.17)
No	320 (336.03)	278 (282.21)	148 (135.93)	82 (73.83)
	487	409	197	107

$$p = \frac{167 + 131 + 49 + 25}{487 + 409 + 197 + 107} = \frac{372}{1200} = 0.31 \qquad \chi^2_{0.05} = 7.815$$

$$\chi^2 = \frac{(167 - 150.97)^2}{150.97} + \frac{(131 - 126.79)^2}{126.79} + \frac{(49 - 61.07)^2}{61.07} + \frac{(25 - 33.17)^2}{33.17}$$

$$+ \frac{(320 - 336.03)^2}{336.03} + \frac{(278 - 282.21)^2}{282.21} + \frac{(148 - 135.93)^2}{135.93} + \frac{(82 - 73.83)^2}{73.83}$$

$$= 1.7021 + 0.1398 + 2.3855 + 2.0123 + 0.7647 + 0.0628 + 1.0718 + 0.9041 = 9.0431$$

Reject null hypothesis. There is a significant difference between the corresponding proportion of women in the various age groups who apply sunscreen.

5.

	Almost every day	Once a week	Once a month	Almost never
Yes	98 (107.4972)	84 (84.6386)	69 (62.3978)	27 (23.4764)
No	76 (66.5028)	53 (52.3614)	32 (38.6022)	11 (14.5236)
	174	137	101	38

$$p = \frac{98 + 84 + 69 + 27}{174 + 137 + 101 + 38} = \frac{278}{450} = 0.6178 \qquad \chi^2_{0.05} = 7.815$$

$$\chi^2 = \frac{(98 - 107.4972)^2}{107.4972} + \frac{(84 - 84.6386)^2}{84.6386} + \frac{(69 - 62.3978)^2}{62.3978} + \frac{(27 - 23.4764)^2}{23.4764}$$

$$+ \frac{(76 - 66.5028)^2}{66.5028} + \frac{(53 - 52.3614)^2}{52.3614} + \frac{(32 - 38.6022)^2}{38.6022} + \frac{(11 - 14.5236)^2}{14.5236}$$

$$= 0.839 + 0.005 + 0.699 + 0.529 + 1.356 + 0.008 + 1.129 + 0.855 = 5.420$$

Do not reject null hypothesis. There is no significant difference in the percentage of constituents who are in favor of the new legislation, regardless of frequency of visits.

9.

	Brighton	Beachgate	Bayview	Shorefront
Yes	129 (144.41)	137 (137.61)	164 (150.89)	158 (155.09)
No	317 (301.59)	288 (287.39)	302 (315.11)	321 (323.91)
	446	425	466	479

$$p = \frac{129 + 137 + 164 + 158}{446 + 425 + 466 + 479} = \frac{588}{1816} = 0.3237885 \qquad \chi^2_{0.05} = 7.815$$

$$\chi^2 = \frac{(129 - 144.41)^2}{144.41} + \frac{(137 - 137.61)^2}{137.61} + \frac{(164 - 150.89)^2}{150.89} + \frac{(158 - 155.09)^2}{155.09}$$

$$+ \frac{(317 - 301.59)^2}{301.59} + \frac{(288 - 287.39)^2}{287.39} + \frac{(302 - 315.11)^2}{315.11} + \frac{(321 - 323.91)^2}{323.91}$$

$$= 1.644 + 0.003 + 1.139 + 0.055 + 0.787 + 0.001 + 0.545 + 0.026 = 4.201$$

Do not reject null hypothesis. There is no significant difference in the proportion of residents who would allow new jails and/or drug rehabilitation centers to be built in their neighborhoods.

13.

		Male	Female
Left-Handed?	Yes	21 (22.38)	13 (11.62)
	No	137 (135.62)	69 (70.38)
		158	82

$$p = \frac{21 + 13}{158 + 82} = \frac{34}{240} = 0.141666667 \qquad \chi^2_{0.01} = 6.635$$

$$\chi^2 = \frac{(21 - 22.38)^2}{22.38} + \frac{(13 - 11.62)^2}{11.62} + \frac{(137 - 135.62)^2}{135.62} + \frac{(69 - 70.38)^2}{70.38}$$

$$= 0.085 + 0.165 + 0.014 + 0.027 = 0.292$$

Do not reject null hypothesis. There is no significant difference between the corresponding proportion of male or female visitors who are left-handed.

ANSWERS TO EXERCISES FOR SECTION 12.3

1.

Observed	58	89	61	37
Expected	61.25	61.25	61.25	61.25

$$\text{Expected} = \frac{58 + 89 + 61 + 37}{4} = 61.25 \qquad \chi^2_{0.05} = 7.815$$

$$\chi^2 = \frac{(58 - 61.25)^2}{61.25} + \frac{(89 - 61.25)^2}{61.25} + \frac{(61 - 61.25)^2}{61.25} + \frac{(37 - 61.25)^2}{61.25}$$

$$= 0.1724 + 12.5724 + 0.0010 + 9.6010 = 22.3468$$

Reject null hypothesis. The proportion of babies delivered during the various seasons is not independent of the season of the year.

5.

Observed	68	52	63	71	75
Expected	65.8	65.8	65.8	65.8	65.8

$$\text{Expected} = \frac{329}{5} = 65.8 \qquad \chi^2_{0.05} = 9.488$$

$$\chi^2 = \frac{(68 - 65.8)^2}{65.8} + \frac{(52 - 65.8)^2}{65.8} + \frac{(63 - 65.8)^2}{65.8} + \frac{(71 - 65.8)^2}{65.8}$$

$$+ \frac{(75 - 65.8)^2}{65.8} = 0.0736 + 2.8942 + 0.1191 + 0.4109 + 1.2863 = 4.7841$$

Do not reject null hypothesis that people in some age-groups are more likely to

be afraid of heights than people in other age-groups.

9.

$$7x + 5x + 4x + 3x + x = 275 + 261 + 184 + 168 + 112$$

$$20x = 1000 \qquad \chi^2_{0.05} = 9.488$$

$$x = 50$$

$7x = 350$
$5x = 250$
$4x = 200$
$3x = 150$
$x = 50$

Observed	275	261	184	168	112
Expected	350	250	200	150	50

ANSWERS TO EXERCISES FOR SECTION 12.3

$$\chi^2 = \frac{(275 - 350)^2}{350} + \frac{(261 - 250)^2}{250} + \frac{(184 - 200)^2}{200} + \frac{(168 - 150)^2}{150} + \frac{(112 - 50)^2}{50}$$

$$= 16.0714 + 0.4840 + 1.2800 + 2.1600 + 76.88 = 96.8754$$

Reject null hypothesis. The proportions of the drinks dispensed are not in the

ratio 7:5:4:3:1

ANSWERS TO EXERCISES FOR SECTION 12.4

1.

	In Favor of Proposal	Against Proposal	Total
East	184 (142.1752)	214 (255.8248)	398
Midwest	130 (177.8976)	368 (320.1024)	498
South	148 (155.0353)	286 (278.9647)	434
Far West	166 (152.8919)	262 (275.1081)	428
Total	628	1130	1758

$$df = (2 - 1)(4 - 1) = 3 \qquad \chi^2_{0.05} = 7.815$$

$$\chi^2 = \frac{(184 - 142.1752)^2}{142.1752} + \frac{(214 - 255.8248)^2}{255.8248} + \frac{(130 - 177.8976)^2}{177.8976}$$

$$+ \frac{(368 - 320.1024)^2}{320.1024} + \frac{(148 - 155.0353)^2}{155.0353} + \frac{(286 - 278.9647)^2}{278.9647}$$

$$+ \frac{(166 - 152.8919)^2}{152.8919} + \frac{(262 - 275.1081)^2}{275.1081} = 12.3039 + 6.8379 + 12.8961$$

$$+ 7.1670 + 0.3193 + 0.1774 + 1.1238 + 0.6246 = 41.450$$

Reject null hypothesis. The region in which a respondent lives is not independent of the resident's opinion on legalizing mercy killings.

5.

	I	II	III	IV	V	Total
Mon	28 (23.5121)	22 (23.99)	26 (25.19)	22 (25.19)	21 (21.11)	119
Tues	25 (26.6734)	27 (27.22)	29 (28.58)	31 (28.58)	23 (23.95)	135
Wed	24 (24.3024)	28 (24.80)	25 (26.04)	26 (26.04)	20 (21.82)	123
Thurs	21 (23.5121)	23 (23.99)	25 (25.19)	26 (25.19)	24 (21.11)	119
Total	98	100	105	105	88	196

$$df = (4 - 1)(5 - 1) = 12 \qquad \chi^2_{0.05} = 21.026$$

$$\chi^2 = \frac{(28 - 23.5121)^2}{23.5121} + \frac{(22 - 23.99)^2}{23.99} + \frac{(26 - 25.19)^2}{25.19} + \frac{(22 - 25.19)^2}{25.19}$$

$$+ \frac{(21 - 21.11)^2}{21.11} + \frac{(25 - 26.6734)^2}{26.6734} + \frac{(27 - 27.22)^2}{27.22} + \frac{(29 - 28.58)^2}{28.58}$$

$$+ \frac{(31 - 28.58)^2}{28.58} + \frac{(23 - 23.95)^2}{23.95} + \frac{(24 - 24.3024)^2}{24.3024} + \frac{(28 - 24.80)^2}{24.80}$$

$$+ \frac{(25 - 26.04)^2}{26.04} + \frac{(26 - 26.04)^2}{26.04} + \frac{(20 - 21.82)^2}{21.82} + \frac{(21 - 23.5121)^2}{23.5121}$$

$$+ \frac{(23 - 23.99)^2}{23.99} + \frac{(25 - 25.19)^2}{25.19} + \frac{(26 - 25.19)^2}{25.19} + \frac{(24 - 21.11)^2}{21.11}$$

$$= 0.857 + 0.165 + 0.026 + 0.404 + 0.001 + 0.105 + 0.002$$

$$+ 0.006 + 0.205 + 0.038 + 0.004 + 0.413 + 0.041 + 0.000$$

$$+ 0.152 + 0.268 + 0.041 + 0.001 + 0.026 + 0.395 = 3.151$$

Do not reject null hypothesis that the number of defective items produced on the various production lines is independent of the day of the week.

9.

	0 - 1	2 - 5	6 - 15	16 - 20	Total
Elementary School	91 (90.08)	82 (76.82)	74 (67.35)	40 (52.75)	287
High School	109 (114.87)	91 (97.97)	79 (85.89)	87 (67.27)	366
College	133 (128.05)	111 (109.21)	96 (95.75)	68 (74.99)	408
Total	333	284	249	195	1061

$$df = (3 - 1)(4 - 1) = 6 \qquad \chi^2_{0.05} = 12.592$$

$$\chi^2 = \frac{(91 - 90.08)^2}{90.08} + \frac{(82 - 76.82)^2}{76.82} + \frac{(74 - 67.35)^2}{67.35} + \frac{(40 - 52.75)^2}{52.75}$$

$$+ \frac{(109 - 114.87)^2}{114.87} + \frac{(91 - 97.97)^2}{97.97} + \frac{(79 - 85.89)^2}{85.89} + \frac{(87 - 67.27)^2}{67.27}$$

$$+ \frac{(133 - 128.05)^2}{128.05} + \frac{(111 - 109.21)^2}{109.21} + \frac{(96 - 95.75)^2}{95.75} + \frac{(68 - 74.99)^2}{74.99}$$

$$= 0.009 + 0.349 + 0.656 + 3.081 + 0.300 + 0.496 + 0.553$$

$$+ 5.789 + 0.191 + 0.029 + 0.001 + 0.651 = 12.105$$

Do not reject null hypothesis that the highest educational level attained by at least one of the partners is independent of the number of years that a marriage will last.

ANSWERS TO EXERCISES FOR SECTION 12.5

1. Expected $= \dfrac{212 + 276 + 198 + 246 + 253}{5} = 237$

    MTB > READ THE FOLLOWING IN C1, C2
    DATA > 212   237
    DATA > 276   237
    DATA > 198   237
    DATA > 246   237
    DATA > 253   237
    DATA > END
            5 rows read

    MTB > LET C3= (C1 - C2)**2/C2
    MTB > PRINT C1 C2 C3

Data Display

ROW	C1	C2	C3
1	212	237	2.63713
2	276	237	6.41772
3	198	237	6.41772
4	246	237	0.31477
5	253	237	1.08017

    MTB > SUM C3

Column Sum

Sum of C3   =   16.8945

Reject null hypothesis. The number of calls received is not independent of the day of the week.

1. Since the null hypothesis is rejected only when the value of the test statistic is too large, the rejection region is always on the right, i.e., it is always right-tailed.

ANSWERS TO CHAPTER TEST - (PAGES 689 - 695)

1.

	Satisfied	Not Satisfied	No Opinion	Total
Under 30	226 (197.63)	182 (204.97)	123 (128.39)	531
30 - 50	117 (133.62)	153 (138.58)	89 (86.80)	359
Over 50	88 (99.75)	112 (103.45)	68 (64.80)	268
Total	431	447	280	1158

$$\frac{(280)(268)}{1158} = 64.80 \qquad \text{Choice (c)}$$

5. Choice (b) $\quad \chi^2_{0.05} = 9.488$

9. $df = 2 - 1 = 1 \quad$ Choice (a)

13.

	Lower Class	Middle Class	Upper Class	Total
Yes	184 (177.57)	167 (163.37)	132 (142.06)	483
No	66 (72.43)	63 (66.63)	68 (57.94)	197
Total	250	230	200	680

$\chi^2_{0.05} = 5.991$

$df = 2$

$$\chi^2 = \frac{(184 - 177.57)^2}{177.57} + \frac{(167 - 163.37)^2}{163.37} + \frac{(132 - 142.06)^2}{142.06}$$

$$+ \frac{(66 - 72.43)^2}{72.43} + \frac{(63 - 66.63)^2}{66.63} + \frac{(68 - 57.94)^2}{57.94}$$

$$= 0.233 + 0.081 + 0.712 + 0.570 + 0.198 + 1.746 = 3.5460$$

Do not reject null hypothesis that the proportion of people in favor of legalizing gambling casinos is the same for all socioeconomic classes.

17.

	Crime-Free Neighbor-hood	Many Play-grounds	Schools Nearby	Shopping Neaby	Parking	Total
Male	178 (185.91)	87 (93.45)	78 (74.07)	57 (62.14)	112 (96.43)	512
Female	196 (188.09)	101 (94.55)	71 (74.93)	68 (62.86)	82 (97.57)	518
Total	374	188	149	125	194	1030

$df = 4$

$\chi^2_{0.05} = 9.488$

$$\chi^2 = \frac{(178 - 185.91)^2}{185.91} + \frac{(87 - 93.45)^2}{93.45} + \frac{(78 - 74.07)^2}{74.07}$$

$$+ \frac{(57 - 62.14)^2}{62.14} + \frac{(112 - 96.43)^2}{96.43} + \frac{(196 - 188.09)^2}{188.09}$$

$$+ \frac{(101 - 94.55)^2}{94.55} + \frac{(71 - 74.93)^2}{74.93} + \frac{(68 - 62.86)^2}{62.86}$$

$$+ \frac{(82 - 97.57)^2}{97.57} = 0.337 + 0.446 + 0.209 + 0.425 + 2.512$$

$$+ 0.333 + 0.440 + 0.207 + 0.420 + 2.483 = 7.810$$

Do not reject null hypothesis that the gender of the respondent is independent of the factor considered important by a potential buyer.

21. $89 + 38 + 42 + 11 + 23 + 75 + 151 = 429$

Sunday	= 0.25(429) = 107.25	Thursday = 0.05(429) = 21.45
Monday	= 0.11(429) = 47.19	Friday = 0.19(429) = 81.51
Tuesday	= 0.09(429) = 38.61	Saturday = 0.28(429) = 120.12
Wednesday	= 0.03(429) = 12.87	

Observed	89	38	42	11	23	75	151
Expected	107.25	47.19	38.61	12.87	21.45	81.51	120.12

$$\chi^2_{0.05} = 12.592$$

$$\chi^2 = \frac{(89 - 107.25)^2}{107.25} + \frac{(38 - 47.19)^2}{47.19} + \frac{(42 - 38.61)^2}{38.61}$$

$$+ \frac{(11 - 12.87)^2}{12.87} + \frac{(23 - 21.45)^2}{21.45} + \frac{(75 - 81.51)^2}{81.51} + \frac{(151 - 120.12)^2}{120.12}$$

$$= 3.1055 + 1.7897 + 0.2976 + 0.2717 + 0.1120 + 0.5199 + 7.9385 = 14.0349$$

Reject null hypothesis. The given percentages are incorrect.

25.

Observed	323	99	95	73
Expected	318	109	89	74

$$\chi^2_{0.05} = 7.815$$

$$\chi^2 = \frac{(323 - 318)^2}{318} + \frac{(99 - 109)^2}{109} + \frac{(95 - 89)^2}{89} + \frac{(73 - 74)^2}{74}$$

$$= 0.0786 + 0.9174 + 0.4045 + 0.0135 = 1.4140$$

Do not reject null hypothesis that the data are consistent with the theory.

1.

	Ignition Shutoff	Steering Wheel Lock	Burglar Alarm	Total
Compact	8 (9.09)	7 (8.18)	15 (12.73)	30
Intermediate	3 (4.85)	8 (4.36)	5 (6.79)	16
Large	9 (6.06)	3 (5.45)	8 (8.48)	20
Total	20	18	28	66

$$df = 4 \qquad \chi^2_{0.05} = 9.488$$

$$\chi^2 = \frac{(8 - 9.09)^2}{9.09} + \frac{(7 - 8.18)^2}{8.18} + \frac{(15 - 12.73)^2}{12.73}$$

$$+ \frac{(3 - 4.85)^2}{4.85} + \frac{(8 - 4.36)^2}{4.36} + \frac{(5 - 6.79)^2}{6.79}$$

$$+ \frac{(9 - 6.06)^2}{6.06} + \frac{(3 - 5.45)^2}{5.45} + \frac{(8 - 8.48)^2}{8.48}$$

$$= 0.131 + 0.171 + 0.406 + 0.705 + 3.030$$

$$+ 0.471 + 1.426 + 1.105 + 0.028 = 7.471$$

5. a)

	Prof. Brier	Prof. Silvernail	Total
Pass	32 (31.38)	38 (38.62)	70
Fail	7 (7.62)	10 (9.38)	17
Total	39	48	87

$$df = 1 \qquad \chi^2_{0.05} = 3.841 \qquad p = \frac{70}{87} = 0.8046$$

$$\chi^2 = \frac{(32 - 31.38)^2}{31.38} + \frac{(38 - 38.62)^2}{38.62} + \frac{(7 - 7.62)^2}{7.62} + \frac{(10 - 9.38)^2}{9.38}$$

$$= 0.012 + 0.010 + 0.051 + 0.041 = 0.114$$

Do not reject null hypothesis.

b)

$$p_1 = \frac{32}{39} \qquad q_1 = \frac{7}{39} \qquad p_2 = \frac{38}{48} \qquad q_2 = \frac{10}{48}$$

$$p = \frac{x_1 + x_2}{n_1 + n_2} = \frac{32 + 38}{39 + 48} = \frac{70}{87} \qquad q = \frac{17}{87}$$

$$\sigma_{p_1 - p_2} \approx \sqrt{pq\left(\frac{1}{n_1} + \frac{1}{n_2}\right)} \approx \sqrt{\left(\frac{70}{87}\right)\left(\frac{17}{87}\right)\left(\frac{1}{39} + \frac{10}{48}\right)} = 0.085479$$

$$z = \frac{p_1 - p_2}{\sigma_{p_1 - p_2}} = \frac{\frac{32}{39} - \frac{38}{48}}{0.085479} = 0.33746$$

138

Do not reject null hypothesis.

c) In both cases, we conclude that we should not reject the null hypothesis.

ANSWERS TO CASE STUDY - (PAGE 697)

1.

	20-29	30-39	40-49	Over 50	Total
Yes	320 (336.03)	278 (282.21)	148 (135.93)	82 (73.83)	828
No	167 (150.97)	131 (126.79)	49 (61.07)	25 (33.17)	372
Total	487	409	197	107	1200

$$df = 3 \qquad \chi^2_{0.05} = 7.815$$

$$\chi^2 = \frac{(320 - 336.03)^2}{336.03} + \frac{(278 - 282.21)^2}{282.21} + \frac{(148 - 135.93)^2}{135.93}$$

$$+ \frac{(82 - 73.83)^2}{73.83} + \frac{(167 - 150.97)^2}{150.97} + \frac{(131 - 126.79)^2}{126.79}$$

$$+ \frac{(49 - 61.07)^2}{61.07} + \frac{(25 - 33.17)^2}{33.17} = 0.765 + 0.063 + 1.072$$

$$+ 0.904 + 1.702 + 0.140 + 2.386 + 2.012 = 9.043$$

Reject null hypothesis. There is a significant difference between the corresponding proportion of vacationers who use sun screen.

CHAPTER 13

ANSWERS TO EXERCISES FOR SECTION 13.4

1.

					Row Total
22	26	28	23	25	124
20	25	32	21	27	125
16	13	19	22	20	90
27	29	21	24	24	125
					464 Grand Total

Cell (1): $\dfrac{124^2 + 125^2 + 90^2 + 125^2}{5} - \dfrac{464^2}{20} = 180.4$

Cell (2): $(22^2 + 26^2 + ... + 24^2 + 24^2) - \left(\dfrac{124^2 + 125^2 + 90^2 + 125^2}{5} \right) = 204.8$

Cell (3): $(22^2 + 26^2 + ... + 24^2 + 24^2) - \dfrac{464^2}{20} = 385.2$

Cell (4): $4 - 1 = 3$

Cell (5): $4(5 - 1) = 16$

Cell (6): $20 - 1 = 19$

Cell (7): $\dfrac{180.4}{3} = 60.1$

Cell (8): $\dfrac{204.8}{16} = 12.8$

Cell (9): $\dfrac{60.1}{12.8} = 4.6953$ or 4.70

Source of Variation	Sum of Squares	Degrees of Freedom	Mean Square	F-ratio
Pills	(1) 180.4	(4) 3	(7) 60.1	(9) 4.70
Error	(2) 204.8	(5) 16	(8) 12.8	
Total	(3) 385.2	(6) 19		

$F_{16}^{3}(0.01) = 5.29$

Do not reject null hypothesis that the average decrease in the blood serum cholesterol levels of people taking each of the different types of pills is the same.

5.

				Row Total
Brand A	22	14	17	53
Brand B	16	18	11	45
Brand C	23	15	18	56
Brand D	19	13	16	48
				202 Grand Total

Cell (1): $\dfrac{53^2 + 45^2 + 56^2 + 48^2}{3} - \dfrac{202^2}{12} = 24.333$

Cell (2): $(22^2 + 14^2 + \ldots + 13^2 + 16^2) - \left(\dfrac{53^2 + 45^2 + 56^2 + 48^2}{3} \right) = 109.37$

141

Cell (3): $(22^2 + 14^2 + \ldots + 13^2 + 16^2) - \dfrac{202^2}{12} = 133.7$

Cell (4): $4 - 1 = 3$

Cell (5): $4(3 - 1) = 8$

Cell (6): $12 - 1 = 11$

Cell (7): $\dfrac{24.333}{3} = 8.1$

Cell (8): $\dfrac{109.37}{8} = 13.7$

Cell (9): $\dfrac{8.1}{13.7} = 0.591$

Source of Variation	Sum of Squares	Degrees of Freedom	Mean Square	F-ratio
Brands	(1) 24.333	(4) 3	(7) 8.1	(9) 0.591
Error	(2) 109.37	(5) 8	(8) 13.7	
Total	(3) 133.7	(6) 11		

$F_8^3 (0.05) = 4.07$

Do not reject null hypothesis.

9.

	Location I	Location II	Location III	Row Total
Autumn	19.4	16.8	16.9	53.1
Winter	18.6	18.3	17.8	54.7
Spring	18.9	17.5	19.9	56.3
Summer	19.3	17.6	18.8	55.7
Column Total	76.2	70.2	73.4	219.8 Grand Total

Cell (1): $\left(\dfrac{53.1^2 + 54.7^2 + 56.3^2 + 55.7^2}{3} \right) - \dfrac{219.8^2}{12} = 1.9567$

Cell (2): $\left(\dfrac{76.2^2 + 70.2^2 + 73.4^2}{4} \right) - \dfrac{219.8^2}{12} = 4.5067$

Cell (3):

$$(19.4^2 + 16.8^2 + \dots + 17.6^2 + 18.8^2) - \left(\dfrac{53.1^2 + 54.7^2 + 56.3^2 + 55.7^2}{3} \right)$$

$$- \left(\dfrac{76.2^2 + 70.2^2 + 73.4^2}{4} \right) + \dfrac{219.8^2}{12} = 4.5933$$

Cell (4): $(19.4^2 + 16.8^2 + \dots + 17.6^2 + 18.8^2) - \dfrac{219.8^2}{12} = 11.0567$

Cell (5): $4 - 1 = 3$

Cell (6): $3 - 1 = 2$

Cell (7): $(4 - 1)(3 - 1) = 6$

Cell (8): $4(3) - 1 = 11$

ANSWERS TO EXERCISES FOR SECTION 13.4

Cell (9): $\dfrac{1.9567}{3} = 0.6522$

Cell (10): $\dfrac{4.5067}{2} = 2.2534$

Cell (11): $\dfrac{4.5933}{6} = 0.7656$

Cell (12): $\dfrac{0.6522}{0.7656} = 0.8519$

Cell (13): $\dfrac{2.2534}{0.7656} = 2.9433$

Source of Variation	Sum of Squares	Degrees of Freedom	Mean Square	F-ratio
Seasons	(1) 1.9567	(5) 3	(9) 0.6522	(12) 0.8519
Locations	(2) 4.5067	(6) 2	(10) 2.2534	(13) 2.9433
Error	(3) 4.5933	(7) 6	(11) 0.7656	
Total	(4) 11.0567	(8) 11		

$F_6^3 (0.01) = 9.78$

Do not reject null hypothesis that the average quantity of dissolved oxygen is the same at all three locations and is the same for all the seasons of the year.

144

13.

						Row Total
Bridge 1	2.9	3.4	2.8	3.9	3.8	16.8
Bridge 2	2.4	2.8	2.9	3.2	2.9	14.2
Bridge 3	3.9	3.9	3.8	3.4	2.5	17.5
Tunnel 1	2.9	2.9	3.4	3.5	3.6	16.3
Tunnel 2	2.7	3.6	2.5	3.2	3.9	15.9
Column Total	14.8	16.6	15.4	17.2	16.7	80.7 Grand Total

Cell (1): $\left(\dfrac{16.8^2 + 14.2^2 + 17.5^2 + 16.3^2 + 15.9^2}{5} \right) - \dfrac{80.7^2}{25} = 1.2264$

Cell (2): $\left(\dfrac{14.8^2 + 16.6^2 + 15.4^2 + 17.2^2 + 16.7^2}{5} \right) - \dfrac{80.7^2}{25} = 0.7984$

Cell (3):

$$(2.9^2 + 3.4^2 + \ldots + 3.2^2 + 3.9^2) - \left(\dfrac{16.8^2 + 14.2^2 + 17.5^2 + 16.3^2 + 15.9^2}{5} \right)$$

$$- \left(\dfrac{14.8^2 + 16.6^2 + 15.4^2 + 17.2^2 + 16.7^2}{5} \right) + \dfrac{80.7^2}{25} = 3.8056$$

Cell (4): $(2.9^2 + 3.4^2 + \ldots + 3.2^2 + 3.9^2) - \dfrac{80.7^2}{25} = 5.8304$

Cell (5): $5 - 1 = 4$

Cell (6): $5 - 1 = 4$

Cell (7): $(5 - 1)(5 - 1) = 16$

145

Cell (8): $5(5) - 1 = 24$

Cell (9): $\dfrac{1.2264}{4} = 0.3066$

Cell (10): $\dfrac{0.7984}{4} = 0.1996$

Cell (11): $\dfrac{3.8056}{16} = 0.23785$

Cell (12): $\dfrac{0.3066}{0.23785} = 1.289$

Cell (13): $\dfrac{0.1996}{0.23785} = 0.8392$

Source of Variation	Sum of Squares	Degrees of Freedom	Mean Square	F-ratio
Locations	(1) 1.2264	(5) 4	(9) 0.3066	(12) 1.289
Days	(2) 0.7984	(6) 4	(10) 0.1996	(13) 0.8392
Error	(3) 3.8056	(7) 16	(11) 0.23785	
Total	(4) 5.8304	(8) 24		

$F_{16}^{4}(0.05) = 3.01$

Do not reject null hypothesis that the average amount of sulfur oxide pollutants is the same at all the bridges and tunnels and is the same for each day of the week.

146

ANSWERS TO EXERCISES FOR SECTION 13.5

1. One-way Analysis of Variance

Analysis of Variance

SOURCE	DF	SS	MS	F	p
Factor	4	472.1	118.0	1.89	0.149
Error	21	1308.3	62.3		
Total	25	1780.3			

Individual 95% CIs For Mean
Based on Pooled StDev

Level	N	Mean	StDev	----+---------+---------+---------+--
C1	6	66.333	12.372	(-------*--------)
C2	5	58.000	5.050	(--------*---------)
C3	7	62.714	7.111	(------*-------)
C4	4	63.750	4.031	(----------*-----------)
C5	4	53.750	5.439	(---------*---------)

```
                              ----+---------+---------+---------+--
Pooled StDev  =     7.893      48.0    56.0     64.0     72.0
```

Do not reject null hypothesis that the average time required to assemble the computer is the same for all of the employees.

ANSWERS TO TESTING YOUR UNDERSTANDING OF THIS CHAPTER'S CONCEPTS - (PAGES 732 - 733)

1. Assumptions for one-way ANOVA

  a) <u>Independent samples</u>: The samples taken from the various populations are

  independent of one another.

  b) <u>Normal populations</u>: The populations from which the samples are obtained are

  (approximately) normally distributed.

ANSWERS TO TESTING YOUR UNDERSTANDING OF THIS CHAPTER'S CONCEPTS - (PAGES 732 - 733)

c) Equal standard deviations: The populations from which the samples are obtained all have the same (often unknown) variance, σ^2.

5. No. We can only conclude that they are not all the same.

ANSWERS TO CHAPTER TEST - (PAGES 733 - 739)

1.

						Row Total
Beach 1	48	39	45	49	42	223
Beach 2	46	38	36	35	48	203
Beach 3	40	32	29	40	35	176
Beach 4	24	59	29	45	42	199
Beach 5	23	35	40	28	41	167
Column Total	181	203	179	197	208	968 Grand Total

Cell (1): $\left(\dfrac{223^2 + 203^2 + 176^2 + 199^2 + 167^2}{5} \right) - \dfrac{968^2}{25} = 399.8$

5. Cell (5): $5(5 - 1) = 20$

9. Cell (9): $\dfrac{99.95}{65.76} = 1.5199$

148

13.

							Row Total
Computer Science	36.0	29.0	34.2	39.0	31.0	32.6		201.8
Nursing	28.7	26.9	30.2	30.8	29.1			145.7
Business	38.3	35.0	33.0	37.1	36.0	30.3	32.4	242.1
Liberal Arts	27.1	23.6	20.0					70.7
Early Childhood Ed.	22.0	26.2	21.0	25.3				94.5
Column Total	152.1	140.7	138.4	132.2	96.1	62.9	32.4	754.8

Cell (1):

$$\left[\frac{(201.8)^2}{6} + \frac{(145.7)^2}{5} + \frac{(242.1)^2}{7} + \frac{(70.7)^2}{3} + \frac{(94.5)^2}{4} \right]$$

$$- \frac{(754.8)^2}{25} = 515.91$$

Cell (2):

$$(36.0^2 + 29.0^2 + \ldots + 21.0^2 + 25.3^2)$$

$$- \left[\frac{(201.8)^2}{6} + \frac{(145.7)^2}{5} + \frac{(242.1)^2}{7} + \frac{(70.7)^2}{3} + \frac{(94.5)^2}{4} \right] = 165.41$$

Cell (3): $(36.0^2 + 29.0^2 + \ldots + 21.0^2 + 25.3^2) - \frac{(754.8)^2}{25} = 681.32$

Cell (4): $5 - 1 = 4$

Cell (5): $1(6 - 1) + 1(5 - 1) + 1(7 - 1) + 1(3 - 1) + 1(4 - 1) = 20$

Cell (6): $25 - 1 = 24$

Cell (7): $\dfrac{515.91}{4} = 128.9775$

Cell (8): $\dfrac{165.41}{20} = 8.2705$

Cell (9): $\dfrac{128.9775}{8.2705} = 15.5949$

Source of Variation	Sum of Squares	Degrees of Freedom	Mean Square	F-ratio
Area of Specialization	(1) 515.91	(4) 4	(7) 128.9775	(9) 15.5949
Error	(2) 165.41	(5) 20	(8) 8.2705	
Total	(3) 681.32	(6) 24		

$F_{20}^{4}(0.05) = 2.87$

Reject null hypothesis. The average starting salary is not the same for all these graduates with the five areas of specialization.

17.

						Row Total
Diet A	140	167	154	174	165	800
Diet B	152	159	146	149	146	752
Diet C	164	138	138	156	149	745
Diet D	158	150	149	159	160	776
						3073

Cell (1): $\left(\dfrac{800^2 + 752^2 + 745^2 + 776^2}{5} \right) - \dfrac{3073^2}{20} = 374.6$

Cell (2): $(140^2 + 167^2 + ... + 159^2 + 161^2) - \left(\dfrac{800^2 + 752^2 + 745^2 + 776^2}{5} \right) = 1450.0$

Cell (3): $(140^2 + 167^2 + ... + 159^2 + 161^2) - \dfrac{3073^2}{20} = 1824.6$

Cell (4): $4 - 1 = 3$

Cell (5): $4(5 - 1) = 16$

Cell (6): $20 - 1 = 19$

Cell (7): $\dfrac{374.6}{3} = 124.8667$

Cell (8): $\dfrac{1450.0}{16} = 90.625$

Cell (9): $\dfrac{124.8667}{90.625} = 1.3778$

Source of Variation	Sum of Squares	Degrees of Freedom	Mean Square	F-ratio
Diets	(1) 374.6	(4) 3	(7) 124.8667	(9) 1.3778
Error	(2) 1450.0	(5) 16	(8) 90.625	
Total	(3) 1824.6	(6) 19		

$F_{16}^{3}(0.01) = 5.29$

Do not reject null hypothesis that the type of diet used significantly affects the average blood pressure level.

21.

	I	II	III	IV	Row Total
A	13	20	16	19	68
B	22	23	29	26	100
C	12	20	13	18	63
Column Total	47	63	58	63	231 Grand Total

Cell (1): $\left(\dfrac{68^2 + 100^2 + 63^2}{4} \right) - \dfrac{231^2}{12} = 201.5$

Cell (2): $\left(\dfrac{47^2 + 63^2 + 58^2 + 63^2}{3} \right) - \dfrac{231^2}{12} = 56.9167$

Cell (3):

$$(13^2 + 20^2 + \ldots + 13^2 + 18^2) - \left(\dfrac{68^2 + 100^2 + 63^2}{4} \right)$$

$$- \left(\dfrac{47^2 + 63^2 + 58^2 + 63^2}{3} \right) + \dfrac{231^2}{12} = 47.8333$$

Cell (4): $(13^2 + 20^2 + \ldots + 13^2 + 18^2) - \dfrac{231^2}{12} = 306.250$

Cell (5): $3 - 1 = 2$

Cell (6): $4 - 1 = 3$

Cell (7): $(3 - 1)(4 - 1) = 6$

Cell (8): $3(4) - 1 = 11$

Cell (9): $\dfrac{201.5}{2} = 100.75$

Cell (10): $\dfrac{56.9167}{3} = 18.9722$

Cell (11): $\dfrac{47.8333}{6} = 7.9722$

Cell (12): $\dfrac{100.75}{7.9722} = 12.6377$

Cell (13): $\dfrac{18.9722}{7.9722} = 2.3798$

Source of Variation	Sum of Squares	Degrees of Freedom	Mean Square	F-ratio
Methods	(1) 201.5	(5) 2	(9) 100.75	(12) 12.6377
Hog Types	(2) 56.9167	(6) 3	(10) 18.9722	(13) 2.3798
Error	(3) 47.8333	(7) 6	(11) 7.9722	
Total	(4) 306.25	(8) 11		

$F_6^2 (0.01) = 10.9$ $F_6^3 (0.01) = 9.78$

153

Reject null hypothesis about the different rations used. There is a significant difference in the ration method used. Do not reject null hypothesis that there is a significant difference in the type of hog used.

25.

									Row Total
Smith	76	61	63	87	96	90	80		553
Carter	72	84	64	93	77	96			486
Jones	53	74	97	88	79	76	85	88	640
Kennedy	88	72	91	69	59				379

2058 Grand Total

Cell (1): $\left(\dfrac{553^2}{7} + \dfrac{486^2}{6} + \dfrac{640^2}{8} + \dfrac{379^2}{5} \right) - \dfrac{2058^2}{26} = 82.59$

Cell (2): $(76^2 + 61^2 + \dots + 69^2 + 59^2) - \left(\dfrac{553^2}{7} + \dfrac{486^2}{6} + \dfrac{640^2}{8} + \dfrac{379^2}{5} \right) = 3775$

Cell (3): $(76^2 + 61^2 + \dots + 69^2 + 59^2) - \dfrac{2058^2}{26} = 3857.59$

Cell (4): $4 - 1 = 3$

Cell (5): $1(7 - 1) + 1(6 - 1) + 1(8 - 1) + 1(5 - 1) = 22$

Cell (6): $26 - 1 = 25$

154

Cell (7): $\dfrac{82.59}{3} = 27.53$

Cell (8): $\dfrac{3775}{22} = 171.5909$

Cell (9): $\dfrac{27.53}{171.5909} = 0.1604$

Source of Variation	Sum of Squares	Degrees of Freedom	Mean Square	F-ratio
Teacher	(1) 82.59	(4) 3	(7) 27.53	(9) 0.1604
Error	(2) 3775	(5) 22	(8) 171.5909	
Total	(3) 3857.59	(6) 25		

$F_{22}^{3}(0.05) = 3.05$

Do not reject null hypothesis that the students in each of these teacher's class are not significantly different in their performance on the calculus I final exam.

ANSWERS TO THINKING CRITICALLY - (PAGE 740)

1. A, B, C, D, and E

ANSWERS TO THINKING CRITICALLY - (PAGE 740)

5.

Analysis of Variance

SOURCE	SS	DF	MS	F	p
SOURCE	103.0412	2	51.5206	0.82	0.326
Factor	565.4700	9	62.83		
Error	668.5112	11			

Level	N	Mean	STDEV
C1	4	11.82	7.641
C2	4	12.69	9.123
C3	4	8.93	8.569

Pooled Stdev

ANSWERS TO CASE STUDY - (PAGE 741)

1.

						Row Total
Algebra	42.75	44.82	43.79	47.53			178.89
Statistics	49.95	47.68	48.50	48.88	49.35		244.36
Calculus	69.95	72.36	71.42	65.23	68.69	74.49	422.14
Liberal Arts	46.23	42.86	45.12				134.21

979.60
Grand Total

ANSWERS TO CASE STUDY - (PAGE 741)

Cell (1): $\left(\dfrac{178.89^2}{4} + \dfrac{244.36^2}{5} + \dfrac{422.14^2}{6} + \dfrac{134.21^2}{3} \right) - \dfrac{979.60^2}{18} = 2335.23$

Cell (2): $(42.75^2 + 44.82^2 + \ldots + 42.86^2 + 45.12^2)$

$- \left(\dfrac{178.89^2}{4} + \dfrac{244.36^2}{5} + \dfrac{422.14^2}{6} + \dfrac{134.21^2}{3} \right) = 72.95$

Cell (3): $(42.75^2 + 44.82^2 + \ldots + 42.86^2 + 45.12^2) - \dfrac{979.60^2}{18} = 2408.19$

Cell (4): $4 - 1 = 3$

Cell (5): $1(4 - 1) + 1(5 - 1) + 1(6 - 1) + 1(3 - 1) = 14$

Cell (6): $18 - 1 = 17$

Cell (7): $\dfrac{2335.23}{3} = 778.41$

Cell (8): $\dfrac{72.95}{14} = 5.2107$

Cell (9): $\dfrac{778.41}{5.2107} = 149.3868$

157

Source of Variation	Sum of Squares	Degrees of Freedom	Mean Square	F-ratio
Textbook Subjects	(1) 2335.23	(4) 3	(7) 778.41	(9) 149.3868
Error	(2) 72.95	(5) 14	(8) 5.2107	
Total	(3) 2408.19	(6) 17		

$F_{14}^{3}(0.05) = 5.56$

Do not reject null hypothesis that the average prices of a college textbook in Algebra and Trigonometry, Statistics, Calculus, and Liberal Arts Mathematics are not significantly different.

CHAPTER 14

ANSWERS TO EXERCISES FOR SECTION 14.3

1. We replace each salary with a plus sign if it is above \$32,073 and with a minus sign if it is below \$32,073. We neglect those salaries that equal \$32,073. We have $- - + -$ $+ + - + - + - -$. Here $n = 12$ and the number of plus signs is 5. From Table 14.1, the test statistic exceeds the chart value of 2, so we do not reject the null hypothesis that the median salary is \$32,073.

5.

Before	After	Sign of Difference
10	9	-
9	8	-
8	8	
6	7	+
5	3	-
9	8	-
9	7	-
9	9	
5	4	-
8	6	-
4	3	-
7	2	-

There are 9 minus signs out of a possible 10 sign changes, so

ANSWERS TO EXERCISES FOR SECTION 14.3

$$\hat{p} = \frac{9}{10} = 0.90$$

$$z = \frac{0.90 - 0.50}{\sqrt{\dfrac{0.50\,(1 - 0.50)}{10}}} = \frac{0.40}{0.158} = 2.53$$

Reject null hypothesis. The number of daily arrests has decreased.

(One can also use Table 14.1 to obtain the same results. Here $n = 10$ and the number of plus signs is 1.)

9.

Before	After	Sign of Difference
12	8	-
6	4	-
10	9	-
8	8	
10	6	-
14	9	-
11	5	-
10	6	-
9	10	+
13	12	-

There are 8 plus signs out of a possible 9 sign changes, so

$$\hat{p} = \frac{8}{9} = 0.8889$$

$$z = \frac{0.8889 - 0.50}{\sqrt{\dfrac{0.50\,(1 - 0.50)}{9}}} = \frac{0.3889}{0.1667} = 2.33$$

Reject null hypothesis. The daily number of felony arrests has decreased.

ANSWERS TO EXERCISES FOR SECTION 14.4

1.

Before	After	Difference, D	Absolute value of Difference, |D|	Rank of |D|	Signed Rank
241	256	-15	15	6	-6
233	249	-16	16	7	-7
262	258	4	4	2.5	2.5
256	265	-9	9	5	-5
271	278	-7	7	4	-4
293	289	4	4	2.5	2.5
312	310	2	2	1	1
270	315	-45	45	9	-9
275	305	-30	30	8	-8
285	342	-57	57	10	-10

Positive sign ranks: $2.5 + 2.5 + 1 = 6$

Negative sign ranks : $(-6) + (-7) + (-5) + (-4) + (-9) + (-8) + (-10) = -49$

We select 6 (the smallest sum). Using $n = 10$ with $\alpha = 0.05$, the critical value is 8 for a two-tailed test. We reject the null hypothesis. We conclude that the number of students completing their lunch has changed significantly after the new caterer was hired.

1.

Result	Group	Rank
29	B	1
36	B	2
37	B	3
39	B	4
40	B	5.5
40	B	5.5
41	B	7.5
41	B	7.5
42	B	9.5
42	B	9.5
43	A	11.5
43	B	11.5
44	B	13.5
44	A	13.5
45	B	15
46	A	16.5
46	B	16.5
47	A	18
48	A	19
49	A	20
51	B	21
52	A	22
53	A	23
54	A	24
56	A	25
57	A	26

$n_1 = 11$ $n_2 = 15$

R = sum of ranks for group A

$= 11.5 + 13.5 + 16.5 + 18 + 19 + 20 + 22 + 23 + 24 + 25 + 26 = 218.5$

ANSWERS TO EXERCISES FOR SECTION 14.5

$$\mu_R = \frac{11(11 + 15 + 1)}{2} = 148.5$$

$$\sigma_R = \sqrt{\frac{11(15)(11 + 15 + 1)}{12}} \approx 19.2678$$

$$z = \frac{218.5 - 148.5}{19.2678} = 3.63$$

Reject null hypothesis. The mean score for both groups is not the same.

5.

Number of Minutes	Group	Rank
45	F	1
47	M	2
49	M	3
54	F	4.5
54	M	4.5
56	F	6.5
56	F	6.5
57	M	8
59	F	9.5
59	M	9.5
61	F	11
62	F	12
63	F	14
63	M	14

63	M	14
65	F	16
66	F	17
67	M	18
69	F	19.5
69	F	19.5
71	M	21
73	F	22
74	M	23

$$n_1 = 13 \quad n_2 = 10$$

R = sum of ranks for group M
$$= 2 + 3 + 4.5 + 8 + 9.5 + 14 + 14 + 18 + 21 + 23 = 117$$

$$\mu_R = \frac{10(10 + 13 + 1)}{2} = 120$$

$$\sigma_R = \sqrt{\frac{10(13)(10 + 13 + 1)}{12}} \approx 16.1245$$

$$z = \frac{117 - 120}{16.1245} = -0.19$$

Do not reject null hypothesis that the average time needed by both groups is the same.

ANSWERS TO EXERCISES FOR SECTION 14.7

1. Let x, y, z and w represent the rankings of recruiters 1, 2, 3, and 4 respectively. We then have

x	y	z	w	$x-y$	$(x-y)^2$	$x-z$	$(x-z)^2$	$y-w$	$(y-w)^2$
1	1	5	5	0	0	-4	16	-4	16
2	5	3	2	-3	9	-3	9	3	9
3	4	4	4	-1	1	-1	1	0	0
4	2	2	3	2	4	2	4	-1	1
5	3	1	1	2	4	4	16	2	4
					18		38		30

a) $R = 1 - \dfrac{6\Sigma(x-y)^2}{5(5^2-1)} = 1 - \dfrac{6(18)}{5(24)} = 1 - 0.9 = 0.1$

$R = 0.1$ is not significant. We cannot conclude that there is a significant correlation between the ratings of recruiters 1 and 2.

b) $R = 1 - \dfrac{6(38)}{5(5^2-1)} = 1 - 1.9 = -0.9$

$R = -0.9$ is not significant. We cannot conclude that there is a significant correlation between the ratings of recruiters 1 and 3.

c) $R = 1 - \dfrac{6(30)}{5(5^2-1)} = 1 - 1.5 = -0.5$

$R = -0.5$ is not significant. We cannot conclude that there is a significant correlation between the ratings of recruiters 2 and 4.

5.

Run	Letters
1	M
2	F
3	MM
4	F
5	MM
6	FFF
7	MM
8	FFF
9	MM
10	FF
11	M

There are 11 runs where we have 10 M's and 10 F's. Here $n_1 = 10$ and $n_2 = 10$

Do not reject the null hypothesis of randomness.

9. The median of the numbers is 18. We then have the following sequence where a =

above median and b = below median. b b a b b b a a a a b a a a b b b

Run	Letters
1	bb
2	a
3	bbb
4	aaaa
5	b
6	aaa
7	bbb

There are 7 runs where we have 8 a's and 9 b's. Here $n_1 = 8$ and $n_2 = 9$

Do not reject null hypothesis of randomness.

1.

Number of Summonses	Police Officer	Rank
8	Walsh	1
10	Nuzzo	2
11	Smith	3
12	Eskey	4
13	Kien	5
14	Smith	6
15	Smith	7
16	Eskey	8
17	Walsh	9
18	Kien	10
19	Walsh	11
20	Nuzzo	12
21	Nuzzo	13
22	Smith	14
23	Kien	15
24	Nuzzo	16
25	Eskey	17
26	Smith	18
27	Walsh	19
28	Kien	20
29	Eskey	21
30	Kien	22
31	Walsh	23
32	Eskey	24
33	Nuzzo	25

Sum of rankings for Smith $R_1 = 3 + 6 + 7 + 14 + 18 = 48$

Sum of rankings for Eskey $R_2 = 4 + 8 + 17 + 21 + 24 = 74$

Sum of rankings for Kien $R_3 = 5 + 10 + 15 + 20 + 22 = 72$

Sum of rankings for Walsh $R_4 = 1 + 9 + 11 + 19 + 23 = 63$

Sum of rankings for Nuzzo $R_5 = 2 + 12 + 13 + 16 + 25 = 68$

Test statistic

$$= \frac{12}{25(25 + 1)} \left[\frac{48^2}{5} + \frac{74^2}{5} + \frac{72^2}{5} + \frac{63^2}{5} + \frac{68^2}{5} \right] - 3(25 + 1) = 1.5951$$

χ^2 with 4 degrees of freedom $= 13.277$. We do not reject the null hypothesis. The data does not indicate that there is a significant difference in the average number of summonses issued by each of these police officers.

1. MTB > READ NON AIR-CONDITION INTO C1, AIR-CONDITION INTO C2
   DATA > 77   80
   DATA > 39   65
   DATA > 85   76
   DATA > 84   83
   DATA > 35   44
   DATA > 13   29
   DATA > 68   59
   DATA > 86   86
   DATA > 56   79
   DATA > 79   84
   DATA > 81   72
   DATA > 80   73
   DATA > 65   76
   DATA > END
          13 rows read.

   MTB > RANK THE VALUES IN C1, PUT RANKS IN C3
   MTB > RANK THE VALUES IN C2, PUT RANKS IN C4
   MTB > PRINT C1-C4

ANSWERS TO EXERCISES FOR SECTION 14.10

Data Display

Row	C1	C2	C3	C4
1	77	80	7	10.0
2	39	65	3	4.0
3	85	76	12	7.5
4	84	83	11	11.0
5	35	44	2	2.0
6	13	29	1	1.0
7	68	59	6	3.0
8	86	86	13	13.0
9	56	79	4	9.0
10	79	84	8	12.0
11	81	72	10	5.0
12	80	73	9	6.0
13	65	76	5	7.5

MTB > CORRELATION COEFFICIENT BETWEEN RANKS IN C3 AND C4

Correlation (Pearson)

Correlation of C3 and C4 = 0.669

Reject null hypothesis. An air-conditioned room on a humid day significantly affects the performance on this literacy test.

ANSWERS TO TESTING YOUR UNDERSTANDING OF THIS CHAPTER'S CONCEPTS - (PAGE 796)

1. Yes. In the one-tailed test, the alternate hypothesis is that the probability distribution for population A is shifted to the right of that for B. In the two-tailed test, the alternate hypothesis is that the probability distribution for population A is shifted to the left or to the right of that for B.

ANSWERS TO CHAPTER TEST - (PAGES 796 - 802)

1. The median of the numbers is 30. We then have the following sequence where a = above median and b = below median.   b b a b a b a a a a a b b b a a b b

Run	Letters
1	bb
2	a
3	b
4	a
5	b
6	aaaaa
7	bbb
8	aa
9	bb

There are 9 runs where we have 9 a's and 9 b's.

Here $n_1 = 9$ and $n_2 = 9$

Do not reject null hypothesis of randomness.   Choice (a)

5.

Before	After	Sign of Difference
132	128	-
147	146	-
153	154	+
179	179	
182	169	-
158	150	-
147	149	+
180	175	-
193	185	-
175	170	-

There are 7 minus signs out of a possible 9 sign changes, so

$$\hat{p} = \frac{7}{9} = 0.7778$$

$$z = \frac{0.7778 - 0.50}{\sqrt{\dfrac{0.50\,(1 - 0.50)}{9}}} = \frac{0.2778}{0.1667} = 1.67$$

Reject null hypothesis.   Choice (a)

9. The median of the numbers is 30. We then have the following sequence where **a** =

above median and b = below median.   a b b a a a a a a a b b b b b b a a b b b a b a

Run	Letters
1	a
2	bb
3	aaaaaaa
4	bbbbbb
5	aa
6	bbb
7	a
8	b
9	a

There are 9 runs where we have 12 a's and 12 b's.

Here $n_1 = 12$ and $n_2 = 12$

Do not reject null hypothesis of randomness.

13.

Before	After	Sign of Difference
132	130	-
147	140	-
120	123	+
131	139	+
99	99	
84	81	-
103	104	+
109	100	-
123	110	-
148	128	-

There are 6 minus signs out of a possible 9 sign changes, so

$$\hat{p} = \frac{6}{9} = 0.6667$$

$$z = \frac{0.6667 - 0.50}{\sqrt{\dfrac{0.50\,(1 - 0.50)}{9}}} = \frac{0.1667}{0.1667} = 1$$

Do not reject null hypothesis that there is a significant difference in the triglyceride level of a person as a result of using this pill.

17.

Number of Practice Sessions	Group	Rank
19	A	1
20	A	2
21	A	3
24	B	4
25	A	5
26	B	6
27	A	7
28	B	8
29	B	9
30	A	10
31	A	11
32	B	12
33	B	13
34	A	14
35	B	15
36	B	16
37	B	17
38	A	18
39	B	19
40	A	20
41	A	21
42	B	22
43	B	23

Sum of rankings for group A

$$R_1 = 1 + 2 + 3 + 5 + 7 + + 10 + 11 + 14 + 18 + 20 + 21 = 112$$

Sum of rankings for Group B

$$R_2 = 4 + 6 + 8 + 9 + 12 + 13 + 15 + 16 + 17 + 19 + 22 + 23 = 164$$

Test statistic

$$= \frac{12}{23 (23 + 1)} \left[\frac{112^2}{11} + \frac{164^2}{12} \right] - 3 (23 + 1) = 1.515$$

We do not reject the null hypothesis that the mean number of practice sessions is significantly different for both groups.

21.

Before	After	Sign of Difference
19	12	-
15	14	-
17	17	
20	21	+
23	21	-
16	19	+
18	18	
10	10	
16	9	-
14	12	-

There are 5 minus signs out of a possible 7 sign changes, so

$$\hat{p} = \frac{5}{7} = 0.7143$$

$$z = \frac{0.7143 - 0.50}{\sqrt{\dfrac{0.50\,(1 - 0.50)}{7}}} = \frac{0.2143}{0.1890} = 1.13$$

Do not reject null hypothesis. The two-way radios have no significant effect on the number of robberies of cab drivers.

ANSWERS TO THINKING CRITICALLY - (PAGES 802 - 803)

1. $R_a = 12$ $n_a = 13$ $n_b = 12$

$$z = \frac{12 - \left(\dfrac{2\,(13)\,(12)}{13 + 13} + 1 \right)}{\sqrt{\dfrac{2\,(13)\,(12)\,(2 \cdot 13 \cdot 12 - 13 - 12)}{(13 + 12)^2\,(13 + 12 - 1)}}} = \frac{-1.48}{2.44} = -0.61$$

Do not reject null hypothesis.

1.

Before	After	Sign of Difference
126	132	+
134	125	-
176	175	-
145	164	+
193	195	+
153	162	+
137	146	+
117	125	+
112	113	+
162	170	+

There are 8 plus signs out of a possible 10 sign changes, so

$$\hat{p} = \frac{8}{10} = 0.8$$

$$z = \frac{0.8 - 0.50}{\sqrt{\dfrac{0.50\,(1 - 0.50)}{10}}} = \frac{0.30}{0.1581} = 1.898$$

Reject null hypothesis. The asthma medication significantly affects a person's weight.